Printed at the Mathematical Centre, 413 Kruislaan, Amsterdam.

The Mathematical Centre, founded the 11-th of February 1946, is a non-profit institution aiming at the promotion of pure mathematics and its applications. It is sponsored by the Netherlands Government through the Netherlands Organization for the Advancement of Pure Research (Z.W.O.).

MATHEMATICAL CENTRE TRACTS 124

CENSORING AND STOCHASTIC INTEGRALS

R.D. GILL

MATHEMATISCH CENTRUM AMSTERDAM 1980

1980 Mathematics subject classification: Primary: 62G05, 62G10, 62G15, 62G20
Secondary: 62M99, 62N05, 62P10

ISBN 90 6196 197 1

ACKNOWLEDGEMENTS

At the completion of this monograph I wish to thank Prof.Dr. J. Oosterhoff from whom I have received so much advice and constructive criticism, Prof.Dr. C.L. Scheffer and Dr. P.C. Sander for their useful comments and Dr. P. Groeneboom who introduced me to this subject.

Moreover I thank the Mathematical Centre for the opportunity to publish this monograph in their series Mathematical Centre Tracts and all those at the Mathematical Centre who have contributed to its realization.

CONTENTS

CHAPTER 1

PRELIMINARIES

1.1. Introduction

Censoring occurs both in industrial life-testing (i.e. investigation of the distribution of the lifetime of manufactured components or complete systems) and in medical trials and biological experiments (e.g. on carcinogens). So terms synonymous to a "censored observation" are a "withdrawal", a "loss", or a "death due to a competing risk"; while an "uncensored observation" might be a "failure", a "relapse", or a "death from the cause under study". More detailed examples are given in Section 3.1.

Formally, in all these situations one is interested in the distribution or distributions of n independent positive random variables $X_1,\ldots,X_n$. However one is only in a position to observe $(\tilde{X}_1,\delta_1),\ldots,(\tilde{X}_n,\delta_n)$ where the δ_j's are indicator random variables (i.e. take the values zero or one only) such that δ_j takes the value 1 if observation j is uncensored, in which case $\tilde{X}_j$ takes the same value as X_j. On the other hand, if δ_j takes the value 0, observation j is censored and we only know that X_j takes a value larger than $\tilde{X}_j$.

In all the situations outlined above, time and random phenomena occurring in time play an essential role. It is our thesis that the same is true of the mathematics of the situation: in other words, it pays to study the statistical problems of interest in terms of the theory of stochastic processes.

This possibility of a new and fruitful application of probability theory to the statistics of censored data was exploited by O.O. Aalen in his thesis, AALEN (1976), and later articles, especially AALEN (1977) and (1978). In particular he made use of the theory of stochastic integrals as developed by the Strasbourg school of probabilists (see MEYER (1976) or JACOD (1979) for recent and complete accounts of the theory) together with the theory of counting processes developed especially in Berkeley by various authors such as BRÉMAUD (1975), DOLIVO (1974), JACOD (1975) and BOEL, VARAIYA & WONG (1975a, 1975b). A general survey of the theory of counting processes is given by BRÉMAUD & JACOD (1977).

We are especially interested in a number of one- and two-sample statistical methods which lend themselves very nicely to a treatment in this framework. In the first case $X_1,\ldots,X_n$ are identically distributed with an unknown distribution function F which one wants to estimate; while in the second case the X_j's fall into two groups, those in group i being identically distributed with distribution function F_i $(i = 1,2)$, and one wants to test the null hypothesis $F_1 = F_2$. The methods considered are approximate and non-parametric: more explicitly, they rely on large-sample results, and do not assume that F, or F_1 and F_2, belong to some parametric family of distributions. In general no truly non-parametric (i.e. distribution-free) methods are possible; at least, not useful ones.

In the first place we consider the product limit estimator of KAPLAN & MEIER (1958), which plays a role for censored data similar to that of the empirical distribution function for uncensored data, and the two-sample test statistics of GEHAN (1965), EFRON (1967) and COX (1972). These test statistics are generalizations of ones originally developed for very special types of censored data; the first two being Wilcoxon-type tests while the last one is of Savage-type. They are the most widely used and applicable non-parametric two-sample tests for use with censored data.

Our plan of attack is as follows. The present chapter closes with a summary of notation and conventions which will be used later without comment. In Chapter 2 we build up an arsenal of results from the theory of stochastic processes in particular concerning stochastic integrals, martingales, counting processes and weak convergence of processes, and the interrelations between these subjects. The returns for using such heavy artillery will be unification and generality. We do not need the full force of many of the original results and so have striven here for simplicity.

Chapter 3 begins with examples of how censored data can arise (we restrict attention till Chapter 6 to so-called right censorship) and then extracts a few key properties of all but one of these examples. A model with these properties underlies the rest of Chapter 3 and all of Chapters 4 and 5. In Section 3.2 we introduce the product limit estimator and in Section 3.3 the three test statistics in terms of the model for censored observations which has been established. By way of illustration of the theory of stochastic integrals, we derive some of the small sample properties of the estimator and the test statistics, the latter being considered as members of a general class of test statistics $\mathcal{K}$. Of particular interest are Theorem 3.2.1 and Proposition 3.2.1, which give linear bounds on the product limit estimator analogous to well known results on the empirical distribution function (see SHORACK & WELLNER (1978) or VAN ZUIJLEN (1978)).

In Chapter 4 we proceed to derive asymptotic results on these statistics. Notations and definitions for this and the following chapter are summarized on pages 53, 54, 55, 58 and 59. As well as giving general results on consistency (Section 4.1) against various types of alternatives and asymptotic normality (Sections 4.2 and 4.3) we specialize to what we call "the general random censorship model" (Example 4.1.1) in which for each j, $\tilde{X}_j = \min(X_j,U_j)$, where $U_1,\ldots,U_n$ are "censoring variables", independent of one another and of the X_j's, and with arbitrary distributions. We also pay special attention to the case when $U_1 = \ldots = U_n = T$ for some "stopping rule" T depending on the observations. The results are derived with a unified approach and at the same time generalize those to be found in the literature. In particular we do not require any of the distribution functions concerned to be continuous, and extend test statistics originally proposed for continuously distributed data for use in the situations where the underlying distribution functions are (partially) discrete.

In Chapter 5 we look at efficiencies when testing against specific alternatives. We develop some new test statistics, also members of $\mathcal{K}$, which are specially suited for testing against particular parametric alternatives. Also we derive test statistics which are consistent when testing against the mere inequality of two distributions.

Finally in Chapter 6 we sketch a number of extensions to the preceding theory. In particular we mention more general forms of censorship than the "right censorship" considered so far, and we pay some attention to the example in Chapter 3 which was not covered by our general model.

1.2. Notation

The following notations will be used without comment in the sequel. Let X be a real-valued function on the set of nonnegative real numbers $\mathbb{R}^+ = [0,\infty)$. If X has finite left hand limits everywhere (we say "X has left hand limits"), then X_- is the function on $\mathbb{R}^+$ defined by $X_-(t) = X(t-)$, $t > 0$, and $X_-(0) = 0$. We define X_+ similarly when X has finite right hand limits everywhere, and define $X(\infty) = \lim_{t\to\infty} X(t)$ if this exists. If X is right continuous with left hand limits then ΔX is the function $X - X_-$. If $\{X_j: j \in J\}$ is some indexed family of functions, we write X_{j-} for $(X_j)_-$, etc.

Suppose Y is a real-valued function on $\mathbb{R}^+$ which is right continuous with left hand limits and is of bounded variation on each bounded subinterval of $\mathbb{R}^+$ (we also say "Y is of locally bounded variation"). Moreover suppose that X is a Lebesgue-measurable real-valued function on $\mathbb{R}^+$ such that $\int_{s\in[0,t]} |X(s)||dY(s)|$ is finite for each $t \in \mathbb{R}^+$ (i.e. "X is locally integrable with respect to Y"). Here the integral is a Lebesgue-Stieltjes integral with respect to the total variation of Y (which assigns mass $|Y(0)|$ to the point zero in line with the convention $Y(0-) = 0$). Then for each t we define

$$(1.2.1) \qquad \int_0^t XdY = \int_{s\in[0,t]} X(s)dY(s),$$

and we denote by $\int XdY$ the *function* taking the value (1.2.1) in the point t. Note that $(\int XdY)(0) = X(0)Y(0)$. We denote by Y_c the continuous part of Y; i.e.

$$(1.2.2) \qquad Y_c(t) = Y(t) - \sum_{s\leq t} \Delta Y(s),$$

where the sum is an absolutely convergent sum of at most countably many nonzero terms.

All the above notations will be extended to stochastic processes in Section 2.1.

$(\Omega,\mathcal{F},P)$ will denote a complete probability space and ω a generic member of Ω. We write $\sigma\{\cdot\}$ for the sub-σ-algebra of $\mathcal{F}$ generated by a family of random variables and use the symbol $\vee$ to denote the σ-algebra generated by a union of σ-algebras. Convergence in probability and in distribution are denoted by $\to_P$ and $\to_{\mathcal{D}}$ respectively. $\mathcal{N}(\mu,\sigma^2)$ is the normal distribution with mean μ and variance σ^2.

The following are some miscellaneous points of notation. χ_A is the indicator variable for the set A. For typographical convenience our notation for an indexed set (i.e. specifying a function) is the same as that for a set itself: we write $\{X(t): t \in [0,\infty)\}$ for the indexed set $\{X(t)\}_{t\in[0,\infty)}$. When dealing with a function of two variables, $(t,\omega) \to X(t,\omega)$, we may write $X(\cdot,\omega)$ for the function of t obtained when ω is fixed. Symbols s,t,u,v,τ are always "time variables" either in $\mathbb{R}^+$ or in $\bar{\mathbb{R}}^+ = [0,\infty]$, while i,j,m,n,r are "index variables" in $\mathbb{N}$. The symbols $\wedge$ and $\vee$ are used to denote minimum and maximum respectively; and # denotes the number of elements in a set. For a real number x, the integral part of x is denoted by [x]. The symbol $\propto$ means "is proportional to". Throughout, we hold to the convention 0/0 = 0.

CHAPTER 2

SOME RESULTS FROM THE THEORY OF STOCHASTIC PROCESSES

2.1. Notation and basic concepts

References for this and the following section are MEYER (1976) or JACOD (1979).

Let $(\Omega,\mathcal{F},P)$ be a fixed complete probability space. A real *stochastic process* $X = \{X(t): t \in [0,\infty)\}$ is a time-indexed family of real-valued random variables. X can therefore also be considered as a function on $[0,\infty) \times \Omega$ and we accordingly write $X(t,\omega)$ for the realized value of the random variable $X(t)$ in the point $\omega \in \Omega$. The *sample paths* or simply *paths* of X are the real-valued functions $X(\cdot,\omega)$ on $[0,\infty)$. If $X(t)$ is integrable for each t, we write $\mathcal{E}X$ for the function $t \to \mathcal{E}(X(t))$. We call X itself *integrable* if $\sup_{t\in[0,\infty)} \mathcal{E}|X(t)|$ is finite; and *square integrable* if X^2 is integrable.

Two processes whose paths are almost surely identical are called *indistinguishable*. When we say that a process for example *is right continuous, has left hand limits*, or *is of finite variation*, we mean (unless explicitly stated otherwise) that almost all of the sample paths have this property. If a process has left hand limits, we can define (up to indistinguishability) a left continuous process X_- such that $X_-(\cdot,\omega) = (X(\cdot,\omega))_-$ for almost all $\omega \in \Omega$. We similarly define processes X_+ and ΔX under the appropriate conditions, at least up to indistinguishability.

In the same way we can define $\int XdY$ and Y_c if almost all the paths of X and Y have the appropriate properties (see (1.2.1) and (1.2.2)). However it is not generally true that this defines stochastic processes, for $\int_{s\in[0,t]}X(s,\cdot)dY(s,\cdot)$ (denoted by $\int_0^t XdY$) and $Y_c(t,\cdot)$ are not necessarily measurable functions on $(\Omega,\mathcal{F})$. In the sequel we often apply the condition that X and Y be *measurable* processes; i.e. as functions of $(t,\omega) \in [0,\infty) \times \Omega$ they should be measurable with respect to the product σ-algebra $\mathcal{B} \otimes \mathcal{F}$, where $\mathcal{B}$ is the Borel σ-algebra on $[0,\infty)$. In particular, processes *all* of whose paths are left continuous or *all* of whose paths are right continuous are measurable. The process $\int|dY|$ is called the *variation* of Y.

Till now the ideas of "past" and "future" have been absent. To introduce them, we suppose that we are given a family $\{F_t: t \in [0,\infty)\}$ of sub-σ-algebras of the complete σ-algebra F such that

(i) $\{F_t\}$ is *increasing*: $F_s \subset F_t$ for all $s < t$,

(ii) $\{F_t\}$ is *right continuous*: $F_s = \bigcap_{t>s} F_t$ for all s,

(iii) $\{F_t\}$ is *complete*: F_0 contains all P-null sets of F.

F_t is to be interpreted as the collection of all events which can occur at or before time t. So (i) expresses the fact that as time evolves, new events may happen. Conditions (ii) and (iii) are technical ones; for us they are completely harmless (see Appendix 2 for some results on how (ii) and (iii) may be verified). We define the σ-algebras $F_{t-} = \bigvee_{s<t} F_s$ and $F_\infty = \bigvee_{t\in[0,\infty)} F_t$.

A collection (Ω,F,P), $\{F_t: t \in [0,\infty)\}$ satisfying the above requirements is called a *stochastic basis*. For the rest of this section we suppose one to be given.

We can now define an *adapted* process X as one such that X(t) is F_t-measurable for each t. A *stopping time* T is an $\overline{\mathbb{R}}^+$-valued random variable such that $\{T\leq t\} \in F_t$ for each t. Interpreting T as the time some random phenomenon occurs, T is a stopping time if at each time instant t one can determine whether or not the phenomenon has yet occurred. The σ-algebra F_T, which can be interpreted as the collection of all events which can take place at or before time T, is defined by

$$F_T = \{A \in F: A \cap \{T\leq t\} \in F_t \; \forall t \in [0,\infty)\}.$$

We next introduce three important classes of processes: martingales, predictable processes, and counting processes. If an adapted process M is *right continuous with left hand limits,* is such that M(t) is integrable for each t, and is such that

$$E(M(t)\,|F_s) = M(s)$$

for each $s < t$, then we call M a *martingale*. If M is a square integrable martingale, then $\lim_{t\to\infty} M(t) = M(\infty)$ exists almost surely, and adjoining F_∞ to the stochastic basis, M is a square integrable martingale on the time set $[0,\infty]$.

A *predictable* process is one measurable with respect to the σ-algebra on $[0,\infty) \times \Omega$ generated by the adapted processes, all of whose paths are left continuous on $(0,\infty)$. So in particular the latter processes and Borel functions of them are predictable; and a deterministic process all of whose paths are equal to a single Borel measurable function is predictable. If H and K are predictable and $\int HdK$ exists, it too is predictable.

A *multivariate counting process* $N = \{N_i: i = 1,\ldots,r\}$ is a finite family of adapted processes N_i such that for almost all $\omega \in \Omega$, the paths of $N_1,\ldots,N_r$ are nondecreasing, right continuous, integer-valued functions, zero at time zero, and with jumps of size +1 only, no two processes jumping at the same time.

Loosely speaking, a martingale is a process without any systematic behaviour in the mean: if M is a martingale then for any s, the process $t \to M(t) - M(s)$, $t \in [s,\infty)$, has zero mean given everything that has happened up to time s. A predictable process is one whose value at time t is fixed given whatever has happened up to but not including time t. This is also true if t is replaced with any stopping time. An r-variate counting process records the occurrences of r types of random phenomena, which cannot occur simultaneously.

A final general concept is that of a process having a certain property *locally*. This is defined by requiring the existence of a so-called *localizing* sequence of stopping times $\{T_n: n \in \mathbb{N}\}$ such that

(i) $T_n \uparrow \infty$ almost surely as $n \to \infty$,

(ii) For each n, the *stopped* process $t \to \chi_{\{T_n>0\}}X(t\wedge T_n)$ has the required property.

If $X(0) = 0$ almost surely, the stopped process above is indistinguishable from the process $t \to X(t\wedge T_n)$, which is MEYER's (1976) definition of stopped process; however our concept of localization is the same. Let us illustrate this important notion by showing that a univariate counting process N is locally bounded (a process is *bounded* if almost all its sample paths are bounded in absolute value by the *same* finite value). For let $T_n = \inf\{t: N(t) \geq n\}$ where the infimum of an empty set is assigned the value $+\infty$. Since the events $\{T_n \leq t\}$ and $\{N(t) \geq n\}$ differ at most by a null set and N is adapted, T_n is a stopping time. Also, $T_n \uparrow \infty$ almost surely. Finally, almost all of the paths of $\chi_{\{T_n>0\}}N(\cdot\wedge T_n)$ are bounded in absolute value by n.

In future we shall generally identify a process with the equivalence class of processes from which it is indistinguishable; this should be particularly borne in mind with statements of equality or uniqueness. It does lead to some anomalies: strictly speaking, only part of the equivalence class of a predictable or a measurable process has these properties.

In the theory of stochastic processes and stochastic integrals, martingales and predictable processes continuously play a complementary role. One instance of this is the following important result on local square integrable martingales. Let M_1 and M_2 be local square integrable martingales. Then there exists a unique predictable process $\langle M_1,M_2\rangle$ whose variation exists and is locally integrable such that $M_1M_2 - \langle M_1,M_2\rangle$ is a local martingale, zero at time zero. If $M_1 = M_2$, $\langle M_1,M_2\rangle$ is in fact non-decreasing. $\langle M_1,M_2\rangle$ is called the *predictable covariation process* of M_1 and M_2. If M_1 and M_2 are in fact square integrable martingales, then $M_1M_2 - \langle M_1,M_2\rangle$ is a martingale on the time interval $[0,\infty]$. Note that $\langle M_1,M_2\rangle$ is right continuous with left hand limits, and that $\langle\cdot,\cdot\rangle$ is symmetric and bilinear.

2.2. Stochastic integrals

In Section 2.1 we saw that under reasonable conditions, the integral of one process with respect to another can be defined in a sensible way and will have all the properties one can reasonably ask of it, such as being a stochastic process itself. The question now arises: what properties of X and Y relative to a given stochastic basis $(\Omega,\mathcal{F},P),\{\mathcal{F}_t: t \in [0,\infty)\}$ carry over to the process $\int XdY$, defined by taking pathwise Lebesgue-Stieltjes integrals of X with respect to Y over the interval $[0,t]$ for each $t \in [0,\infty)$? We already saw that if X and Y are predictable and $\int XdY$ exists, then it is predictable too. It turns out on the other hand that if X is predictable but Y is a *martingale*, then subject to some natural conditions $\int XdY$ is a martingale.

Here we summarize some of the results on this theme, not in the most general form (see MEYER (1976) or JACOD (1979)) but suitable for our purposes.

Let M_1 and M_2 be local square integrable martingales with paths of locally bounded variation, and let H_1 and H_2 be predictable and locally bounded (in particular, H_1 and H_2 have these properties if they are left continuous with right hand limits and are adapted). Then $\int H_1dM_1$ and $\int H_2dM_2$ exist and are local square integrable martingales, and their predictable covariation process satisfies

$$\langle \int H_1dM_1, \int H_2dM_2\rangle = \int H_1H_2d\langle M_1,M_2\rangle.$$

(In fact the requirement that H_i be locally bounded can be relaxed to requiring that $\int H_i dM_i$ exists and $\int H_i^2 d\langle M_i,M_i\rangle$ be locally integrable; however we will hardly ever need this.) If the localizing sequences of stopping times associated with M_1,M_2,H_1 and H_2 are sequences of constants, then the same holds for the localizing sequences associated with $\langle M_1,M_2\rangle$, $\int H_1 dM_1$, etc.; and if the words "local" and "locally" applied to M_1,M_2,H_1 and H_2 can be dropped altogether, the same applies to $\langle M_1,M_2\rangle$, $\int H_1 dM_1$, etc.

We shall make much use of the following corollary of these facts. Let M_1 and M_2 be local square integrable martingales with paths of locally bounded variation, zero at time zero, and let H_1 and H_2 be locally bounded predictable processes. Suppose the localizing sequences of stopping times associated with M_1,M_2, H_1 and H_2 can be taken to be sequences of constants. Then the processes $\int H_1 dM_1$ and $\int H_2 dM_2$ exist and the following equalities between real-valued functions on $[0,\infty)$ hold:

(2.2.1) $$E(\int H_i dM_i) = 0, \quad i = 1,2,$$

(2.2.2) $$E(\int H_1 dM_1 \int H_2 dM_2) = E(\int H_1 H_2 d\langle M_1,M_2\rangle).$$

If the words "local" and "locally" can be dropped altogether, and if $\int H_1 dM_1$ and $\int H_2 dM_2$ are also defined in the point ∞, then the same equalities hold on $[0,\infty]$.

In fact (2.2.1) also holds more generally. Suppose that M is a local martingale (not necessarily locally square integrable) with paths of locally bounded variation, and suppose H is a locally bounded predictable process. Then $\int HdM$ exists and is a local martingale. Now a local martingale is localized by any sequence of stopping times making its variation locally integrable. So if for all t, $E\int_0^t |H||dM| < \infty$, then $\int HdM$ is a martingale. If furthermore $M(0) = 0$ almost surely, then (2.2.1) holds (dropping the index i).

2.3. Counting processes

In this section we show how certain local square integrable martingales are associated with the multivariate counting processes defined in Section 2.1. Recall that these could be interpreted as processes counting the occurrences of a finite number of types of mutually exclusive phenomena. As in Section 2.2 we considerably specialize the general results available; see

BRÉMAUD & JACOD (1977) for a survey of these.

Let $(\Omega,F,P),\{F_t: t \in [0,\infty)\}$ be a fixed stochastic basis and $\{N_i: i = 1,\dots,r\}$ be an r-variate counting process. By MEYER (1976) Theorem I.9, there exist right continuous, nondecreasing, predictable processes A_i, zero at time zero, such that

$$M_i = N_i - A_i \quad i = 1,\dots,r \tag{2.3.1}$$

are local martingales. A_i is called the *compensator* of N_i (and also its "dual predictable projection").

The following result shows that, for each i, M_i is in fact a local square integrable martingale and gives explicit expressions for $\langle M_i,M_j\rangle$. It was proved under the condition that $A_1,\dots,A_r$ are continuous by BOEL, VARAIYA & WONG (1975a); this condition was later removed by ELLIOT (1976), LIPTSER & SHIRYAYEV (1978) and GILL (1978). We give a short proof based on an idea of J. VAN SCHUPPEN in Appendix 1.

THEOREM 2.3.1. *In the situation specified above, each compensator* A_i *satisfies* $0 \le \Delta A_i \le 1$. *The* M_i*'s are local square integrable martingales with*

$$\langle M_i,M_i\rangle = \int (1 - \Delta A_i)dA_i, \tag{2.3.2}$$

$$\langle M_i,M_j\rangle = -\int \Delta A_i dA_j \quad i \neq j, \quad i,j = 1,\dots,r. \tag{2.3.3}$$

The localizing stopping times may everywhere be taken to be any nondecreasing sequence of stopping times $\{T_n\}$, $T_n \to \infty$ *a.s. as* $n \to \infty$, *such that* $E\sum_{i=1}^r N_i(T_n) < \infty$ *for each* $n = 1,2,\dots$ *(here* $N_i(\infty) = \sup_t N_i(t)$*).*

To make use of this result we need to know the processes A_i. We shall make use of the following theorem, adapted from a theorem of MURALI-RAO (1969):

THEOREM 2.3.2. *Let* N *be a univariate counting process and let* $t \in (0,\infty)$ *satisfy* $E(N(t)) < \infty$. *Define*

$$t_{n,i} = i2^{-n}t, \quad n = 1,2,\dots, \quad i = 0,1,\dots,2^n$$

and

$$U_n = \sum_{i=0}^{2^n-1} E(N(t_{n,i+1}) - N(t_{n,i})\,|\,F_{t_{n,i}}), \quad n = 1,2,\dots .$$

Then there exists a subsequence of integers $\{r_n\}$, $r_n \to \infty$ *as* $n \to \infty$, *and a unique random variable* U, *such that for all bounded random variables* X,

$$E(XU_{r_n}) \to E(XU)$$

as $n \to \infty$. *The compensator* A *of* N *satisfies*

$$A(t) = U$$

almost surely.

Note that if $EN(t) = \infty$, one can still apply this theorem to the bounded counting process $N \wedge n$ for each n and take limits; and in the multivariate case, the theorem can be applied to each component in turn. Also it often turns out that the sequence of random variables $\{U_n\}$ is almost surely convergent as $n \to \infty$, so U must be this limit. However the theorem only supplies us with a random variable $U = U_t$ almost surely equal to A(t). To construct A, one should note that the facts: A is right continuous, and $A(t) = U_t$ almost surely for each t, determine A given $\{U_t: t \in [0,\infty)\}$ up to indistinguishability.

Many other theorems can be applied to determine the compensators A_i of a counting process $\{N_i: i = 1,\dots,r\}$. For instance, define (T_n, I_n), $n = 1,2,\dots$ by

$$T_n = \inf\{t: \sum_{i=1}^{r} N_i(t) \geq n\}, \qquad n = 1,2,\dots \tag{2.3.4}$$

and

$$I_n = i \Longleftrightarrow T_n < \infty \text{ and } \Delta N_i(T_n) = 1, \tag{2.3.5}$$

otherwise $I_n = 0$. So T_n is the time of the n-th jump of $\{N_1,\dots,N_r\}$, and if $T_n < \infty$, I_n is the index of the component which then jumps.

Suppose also that

$$F_t = F_0 \vee \sigma\{N_i(s): i = 1,\dots,r;\ s \leq t\}. \tag{2.3.6}$$

(Theorem A.2.1 shows that $\{F_t\}$ is automatically right continuous in this case.) Then Proposition 3.1 of JACOD (1975) shows how the processes $A_1,\dots,A_r$ can be constructed from the conditional distributions of T_{n+1} and I_{n+1} given $F_0, T_1, I_1, \dots, T_n, I_n$ for each n. Conversely, $A_1,\dots,A_r$ in a sense determine the joint distribution of $T_1, I_1, T_2, I_2, \dots$ given F_0 as we shall see presently.

Another theorem by which A_i can be determined is DOLIVO (1974) Theorem 2.5.1 which shows that in certain circumstances $A_i(t)$ may be identified with $\int_0^t \Lambda_i(s)ds$ where

$$\Lambda_i(s) = \lim_{h\downarrow 0} \frac{1}{h} P(N_i(s+h) - N_i(s) \geq 1 \mid \mathcal{F}_s).$$

This result shows that the compensator of a counting process can be interpreted as the integrated or cumulative conditional rate at which it jumps; it can often be used heuristically to suggest what A_i is. In the discrete case where $\mathcal{F}_t = \mathcal{F}_{[t]}$ and N_i only jumps at integer time instants, Theorem 2.3.2 can be applied to show that A_i too is constant between time instants, and that $\Delta A_i(t) = P(\Delta N_i(t) = 1 \mid \mathcal{F}_{t-1})$, $t = 1,2,\ldots$. Again A_i can be interpreted as a cumulative conditional rate for N_i.

A final method for determining A_i is to make use of theorems on uniqueness and existence of processes with a given "intensity process" Λ_i, and then show that the so constructed processes N_i are indeed those one had in mind. Such theorems are given in BOEL, VARAIYA & WONG (1975b), while AALEN (1976) Section 5D illustrates this approach.

We now present two theorems showing that the compensators A_i determine in a sense the probability distribution of the original counting process. The first one is a simplified version of Theorem 5.1 of JACOD (1975):

THEOREM 2.3.3. *Let* $N = \{N_1,\ldots,N_r\}$ *be an r-variate counting process, define* (T_n, I_n), $n = 1,2,\ldots$ *by* (2.3.4) *and* (2.3.5), *and suppose that* $\{\mathcal{F}_t\}$ *is given by* (2.3.6). *Suppose also that* $\sum_{i=1}^r N_i(\infty)$ *is almost surely finite. Let* P' *be another probability measure on* $(\Omega,\mathcal{F})$ *such that* P *and* P' *agree on* $\mathcal{F}_0$ *and are absolutely continuous with respect to one another on* $\mathcal{F}_\infty$. *Suppose* N_i *has compensator* A_i *under* P *and compensator* A_i' *under* P'. *Then for each* i, A_i *and* A_i' *are almost surely absolutely continuous with respect to one another as functions on* $[0,\infty)$, *and on* $\mathcal{F}_\infty$ *we have*

$$\frac{dP'}{dP} = \left(\prod_{n:T_n<\infty} \frac{dA'_{I_n}}{dA_{I_n}}(T_n)\right) \frac{\left(\prod_{s\notin\{T_1,T_2,\ldots\}} (1 - \sum_i \Delta A_i'(s))\right)\exp\left(-\sum_i A'_{ic}(\infty)\right)}{\left(\prod_{s\notin\{T_1,T_2,\ldots\}} (1 - \sum_i \Delta A_i(s))\right)\exp\left(-\sum_i A_{ic}(\infty)\right)}.$$

The final theorem of this section states in effect that if the compensator A of a univariate counting process N is such that for each t, A(t) is determined by the value of N(s), s ≤ t, then the form of A actually determines the probability distribution of the jump times of N. (A multivariate version of the theorem also holds, but we shall not need it.) A proof is given in Appendix 3, in which results of JACOD (1975, 1979) are applied.

THEOREM 2.3.4. *Let* N *be a univariate counting process with compensator* A, *and define* $T_n = \inf\{t: N(t) \geq n\}$, $n = 0,1,\ldots$. *Suppose that outside of a null set of* Ω,

$$A(t) = A(T_n) + f_n(t-T_n;T_1,\ldots,T_n) \qquad \textit{for all } t \in (T_n,T_{n+1}], \quad n = 0,1,\ldots,$$

where f_n $(n = 0,1,\ldots)$ *is a real measurable function on* $(\mathbb{R}^+)^{n+1}$ *such that for* $0 < t_1 < \ldots < t_n$, $f_n(\cdot;t_1,\ldots,t_n)$ *is nondecreasing, right continuous, and zero at time zero. Then the joint probability distribution of* $T_1,T_2,\ldots$ *is uniquely determined by* $f_0,f_1,f_2,\ldots$.

The compensator A of N can be expressed in the form given in Theorem 2.3.4 if for all t

$$\mathcal{F}_t = \mathcal{F}_0 \vee \sigma\{N(s): s \leq t\}$$

and if $\mathcal{F}_0$ is independent of $T_1,T_2,\ldots$ (which is trivially the case if $\mathcal{F}_0$ contains only P-null sets and their complements). For then by JACOD (1975) Proposition 3.1 and Theorem A.2.1,

$$f_n(s;t_1,\ldots,t_n) = \int_0^s \frac{dF_n(u;t_1,\ldots,t_n)}{1 - F_n(u-;t_1,\ldots,t_n)},$$

where F_n is a regular version of the conditional distribution function of $T_{n+1} - T_n$ given $T_1,\ldots,T_n$.

2.4. A martingale central limit theorem and related results

Suppose that for each $n = 1,2,\ldots$ a stochastic basis is given on which r local square integrable martingales Z_i^n, $i = 1,\ldots,r$, are defined. Then for each n, $Z^n = \{Z_i^n: i = 1,\ldots,r\}$ can be considered as a random element of $(D[0,\infty))^r$ where $D[0,\infty)$ is the space of functions on $[0,\infty)$ which are right continuous with finite left hand limits, endowed with the Skorohod topology (see STONE (1963), LINDVALL (1973) or VERVAAT (1972)).

Let A_i^∞, $i = 1,\ldots,r$, be nondecreasing continuous functions on $[0,\infty)$, zero at time zero. It is well known that a random element $Z^\infty = \{Z_i^\infty: i = 1,\ldots,r\}$ of $(D[0,\infty))^r$ can be defined with the following properties: the Z_i^∞'s, $i = 1,\ldots,r$, are independent Gaussian processes with continuous sample paths, zero at time zero, and have zero means, uncorrelated (hence independent) increments, and variance functions A_i^∞, $i = 1,\ldots,r$, i.e.

$$\text{(2.4.1)} \qquad \operatorname{var}(Z_i^\infty(t)) = A_i^\infty(t).$$

In fact the Z_i^∞'s are local square integrable martingales with respect to the natural stochastic basis (let $\mathcal{F}_t^\infty = \sigma\{Z_i^\infty(s): i = 1,\ldots,r,\ s \le t\} \vee \mathcal{N}$, where $\mathcal{N}$ consists of all P-null sets and their complements). We can drop the word "local" if $A_i^\infty(\infty) < \infty$ for each i. Also

$$\text{(2.4.2)} \qquad \langle Z_i^\infty, Z_j^\infty \rangle = \begin{cases} A_i^\infty & i = j \\ 0 & i \ne j. \end{cases}$$

This well known fact has a converse. Suppose processes Z_i^∞, $i = 1,\ldots,r$, are local square integrable martingales with continuous paths such that (2.4.2) holds for given nondecreasing functions A_i^∞, zero at time zero. Then the Z_i^∞'s are r independent Gaussian processes with independent increments and of course (2.4.1) holds; see e.g. MEYER (1971).

This result provides the key idea in the proof of a theorem of REBOLLEDO (1979a), which states that if the jumps of the processes Z_i^n, $i = 1,\ldots,r$, become small in a certain sense as $n \to \infty$, and if $\langle Z_i^n, Z_j^n \rangle(t) \to_P \langle Z_i^\infty, Z_j^\infty \rangle(t)$ as $n \to \infty$ for all i, j and t, then $Z^n \to_{\mathcal{D}} Z^\infty$ as $n \to \infty$ in $(D[0,\infty))^r$. In other words, if in the limit Z^n has the properties which characterize the distribution of Z^∞, then Z^n converges in distribution to Z^∞.

To make the statement concerning the jumps of Z_i^n more precise, let us introduce the concept of an ε-decomposition of r local square integrable martingales $Z_1,\ldots,Z_r$. For $\varepsilon > 0$ let $\bar{Z}_1^\varepsilon,\ldots,\bar{Z}_r^\varepsilon$, $\underline{Z}_1^\varepsilon,\ldots,\underline{Z}_r^\varepsilon$ be local square integrable martingales such that for each i,

$$\text{(2.4.3)} \qquad Z_i = \underline{Z}_i^\varepsilon + \bar{Z}_i^\varepsilon,$$

$$\text{(2.4.4)} \qquad \sup_{t\in[0,\infty)} |\Delta \underline{Z}_i^\varepsilon(t)| \le \varepsilon \text{ almost surely,}$$

(2.4.5) $\bar{Z}_i^\varepsilon$ has paths of locally bounded variation, and for each i and j

$$P(\exists t \in [0,\infty) \text{ such that } \Delta \underline{Z}_i^\varepsilon(t) \neq 0 \text{ and } \Delta \bar{Z}_j^\varepsilon(t) \neq 0) = 0.$$

Then we call $\{\bar{Z}_1^\varepsilon,\ldots,\bar{Z}_r^\varepsilon\}$ the *jump part of an ε-decomposition of* $\{Z_1,\ldots,Z_r\}$. Intuitively speaking, $\{\bar{Z}_1^\varepsilon,\ldots,\bar{Z}_r^\varepsilon\}$ removes completely all the jumps of $\{Z_1,\ldots,Z_r\}$ for which any of the component jumps is greater in absolute value than ε. As an example, let N be a univariate counting process with compensator A, let $M = N - A$, and let H be a locally bounded predictable process. Define $Z = \int H dM$ and $\bar{Z}^\varepsilon = \int H\chi_{\{|H|\geq\varepsilon\}} dM$. Then $\bar{Z}^\varepsilon$ is the jump part of an ε-decomposition of the local square integrable martingale Z.

We now formulate our version of REBOLLEDO's (1979a) Theorem V.I.:

<u>THEOREM 2.4.1</u>. *Let* Z^n, $n = 1,2,\ldots$ *and* Z^∞ *be defined as above and suppose that for each* $\varepsilon > 0$ *and each* $n = 1,2,\ldots$ *an ε-decomposition of* Z^n *exists such that*

(2.4.6) $$\langle \bar{Z}_i^{n\varepsilon}, \bar{Z}_i^{n\varepsilon}\rangle(t) \to_P 0$$

as $n \to \infty$ *for each* i *and* t. *If also*

(2.4.7) $$\langle Z_i^n, Z_j^n\rangle(t) \to_P \begin{cases} A_i^\infty(t) & i = j \\ 0 & i \neq j \end{cases}$$

as $n \to \infty$, *for all* i, j *and* t, *then*

(2.4.8) $$Z^n \to_{\mathcal{D}} Z^\infty$$

as $n \to \infty$ in $(D[0,\infty))^r$. *Furthermore, if* Z_i^n *has paths of locally bounded variation for all* i *and* n, *then*

(2.4.9) $$\sum_{s\leq t} \Delta Z_i^n(s)\, \Delta Z_j^n(s) \to_P \begin{cases} A_i^\infty(t) & i = j \\ 0 & i \neq j \end{cases}$$

as $n \to \infty$ *for all* i, j *and* t.

This theorem is also valid with $[0,\infty)$ replaced everywhere by $[0,\infty]$, noting that on $[0,\infty]$ localizing stopping times T_n, $n = 1,2,\ldots$, should also satisfy $P(T_n = \infty) \to 1$ as $n \to \infty$, and that we now also require $A_i^\infty(\infty) < \infty$, $i = 1,\ldots,r$.

In REBOLLEDO (1979a), the theorem is given for the case $r = 1$ but our version can be obtained from this one by a straightforward application of the Cramèr-Wold device (see REBOLLEDO (1978) Theorem 3.5 for a similar extension). Also the original theorem requires (2.4.7) to hold for the canonical ε-decomposition, which we prefer not to introduce. However the proof of REBOLLEDO (1979b) Lemma 5 part 2 shows that it suffices to assume that any ε-decomposition exists such that (2.4.6) holds.

Recently HELLAND (1980) has given more elementary proofs of REBOLLEDO's theorems, while LIPTSER & SHIRYAYEV (1980) have proved a remarkably general central limit theorem which contains REBOLLEDO's as a special case. However in our applications the conditions become essentially equivalent.

The following result of LENGLART (1977) has at first sight nothing to do with martingale central limit theorems. However it is a major tool in REBOLLEDO's proof of Theorem 2.4.1, and we shall have repeated occasion to use it in conjunction with the previous theorem. A fixed stochastic basis is supposed to be given.

THEOREM 2.4.2. *Let X and Y be adapted, right continuous, nonnegative processes, and suppose also that Y is nondecreasing, zero at time zero, and predictable. Suppose that for all almost surely finite stopping times T, $\mathcal{E}X(T) \le \mathcal{E}Y(T)$. Then for any stopping time T and any $\varepsilon,\eta > 0$,*

$$P(\sup_{s\le T, s<\infty} X(s) \ge \varepsilon) \le \frac{\eta}{\varepsilon} + P(Y(T) > \eta).$$

There are two basic ways in which we will make use of Theorem 2.4.2. Suppose that N is a univariate counting process with compensator A. Suppose that $\mathcal{E}N(\infty) < \infty$ so that by Theorem 2.3.1 $M = N - A$ is a square integrable martingale. Let H be a nonnegative, bounded, predictable process. Then the conditions of Theorem 2.4.2 are satisfied if we take $X = \int HdN$ and $Y = \int HdA$, because $\int HdM$ is a martingale on $[0,\infty]$ and so for any stopping time T, $\mathcal{E}\int_0^T HdM = 0$. Thus for any stopping time T and $\varepsilon,\eta > 0$,

$$P\left(\int_0^T HdN \ge \varepsilon\right) \le \frac{\eta}{\varepsilon} + P\left(\int_0^T HdA > \eta\right).$$

On the other hand, let N, A and H be as above, except that H is not necessarily nonnegative. We have

$$(\int HdM)^2 - \int H^2 d\langle M,M\rangle$$

is a martingale on $[0,\infty]$, and Theorems 2.4.2 and 2.3.1 now yield

$$P\left(\sup_{s\le T, s<\infty}\left(\int_0^s HdM\right)^2 \ge \varepsilon\right) \le \frac{\eta}{\varepsilon} + P\left(\int_0^T H^2(1-\Delta A)dA > \eta\right)$$

$$\le \frac{\eta}{\varepsilon} + P\left(\int_0^T H^2 dA > \eta\right).$$

Let us also point out one link between Theorems 2.4.1 and 2.4.2: the latter can be used to show that condition (2.4.6) implies that for all $t \in [0,\infty)$ and $\varepsilon > 0$,

$$\sup_{[0,t]} |\bar{Z}_i^{n\varepsilon}| \to_P 0 \quad \text{as } n \to \infty.$$

Hence condition (2.4.6) together with (2.4.3) and (2.4.4) can indeed be interpreted as stating that the jumps of Z_i^n disappear as $n \to \infty$.

We now turn to a very different subject. The Skorohod-Dudley theorem (see DUDLEY (1968) Theorem 3, or WICHURA (1970)) can be thought of as providing a converse to the well known result that an almost surely convergent sequence of random variables also converges in distribution. Because almost sure convergence is stronger than convergence in distribution, the theorem often provides a short cut in deriving new convergence in distribution results from old ones.

<u>THEOREM 2.4.3</u>. *Let $Z^\infty, Z^1, Z^2, \ldots$ be random elements taking values in a separable metric space such that $Z^n \to_{\mathcal{D}} Z^\infty$ as $n \to \infty$. Then there exists a probability space with random elements $Z^{\infty'}, Z^{1'}, Z^{2'}, \ldots$ defined on it such that $Z^{\infty'}$ has the same distribution as Z^∞ and $Z^{n'}$ has the same distribution as Z^n, $n = 1,2,\ldots$, and such that $Z^{n'} \to Z^{\infty'}$ almost surely as $n \to \infty$.*

Not surprisingly we shall be applying Theorem 2.4.3 with the separable metric space in question being $D([0,u))$ or $D([0,u])$ for some $u \in (0,\infty]$. Suppose we have shown that $Z^n \to_{\mathcal{D}} Z^\infty$ on $D(I)$ when I is $[0,u)$ or $[0,u]$. We shall of course consider the random elements Z^n and Z^∞ of $D(I)$ as stochastic processes as $t \in I$ varies. Suppose that Z^∞ with probability 1 has continuous sample paths. Then because convergence in the Skorohod topology on a closed interval to a continuous limit is equivalent to convergence in the supremum norm on that interval, Theorem 2.4.3 supplies us with processes $Z^{n'}$ and $Z^{\infty'}$ defined on a single probability space with the same distributions as Z^n and Z^∞ respectively, such that

$$\sup_{[0,t]} |Z^{n'} - Z^{\infty'}| \to 0$$

almost surely as $n \to \infty$ for all $t \in I$ (see VERVAAT (1972) Assumption 1.3.3 and the remarks at the beginning of his Section 1.4).

Note that if Z^∞ is a Gaussian process with expectation zero, independent increments, and variance function $A^\infty(t) = \mathrm{var}(Z^\infty(t)) = \mathrm{cov}(Z^\infty(t), Z^\infty(u))$ if $t \leq u$, then Z^∞ has continuous paths if and only if A^∞ is continuous; in general, Z^∞ only jumps at the jump times of A^∞.

CHAPTER 3

RIGHT CENSORSHIP
AND STOCHASTIC INTEGRALS

3.1. Background

In this section we derive a property common to a number of important models for "n censored observations", where n is considered fixed and the censorship is really "right censorship": only in Chapter 6 will we consider general censorship.

We want to model the situation commonly occurring in medical follow-up trials, industrial life-testing, biological experimentation, and other fields, in which one is interested in certain aspects of the distributions of n independent positive random variables $X_1,\ldots,X_n$, but either deliberately or accidentally is only in a position to observe certain bivariate random variables $(\tilde{X}_1,\delta_1),\ldots,(\tilde{X}_n,\delta_n)$ where for each j, $0 < \tilde{X}_j \le X_j$ and $\delta_j = \chi_{\{X_j=\tilde{X}_j\}}$. If δ_j takes the value 1, the j-th observation is *uncensored* and the observed value of $\tilde{X}_j$ is also the realized value of X_j. However if $\delta_j = 0$, the j-th observation is *censored at time* $\tilde{X}_j$, and one only knows that X_j takes (or would have taken) a value strictly greater than the observed value of $\tilde{X}_j$.

One might be interested in comparing the distribution functions of the X_j's in particular subgroups, or in estimating some characteristics of the distribution functions. However for the time being we do not consider the purpose of the experiment. We start with a number of examples of different situations involving different types of censored data, giving them their traditional names.

EXAMPLE 3.1.1 "(Simple) Type I censorship".
In industrial life-testing, $X_1,\ldots,X_n$ are supposed to be n independent and identically distributed positive random variables, with distribution function F. Often it is thought that $F = F_\theta$, where $\{F_\theta\colon \theta \in \Theta\}$ is some parametrized family of distributions. The random variables X_i represent the lengths

of time that n manufactured components function satisfactorily, each operating from time zero under fixed working conditions. The components are observed up to a fixed time instant $u > 0$, at which time not all components may have "failed". So the data on which e.g. estimation of θ or testing of the hypotheses $F \in \{F_\theta: \theta \in \Theta\}$ is to be based is $(\tilde{X}_j,\delta_j) = (X_j \wedge u, \chi_{\{X_j \le u\}})$, $j = 1,\ldots,n$.

EXAMPLE 3.1.2 "(Simple) Type II censorship".
In the situation of Example 3.1.1, instead of terminating the experiment at the fixed time u, it is terminated at the time of the r-th observed failure for some fixed $r \le n$. So if $X_{(1)} \le \ldots \le X_{(n)}$ are the order statistics of $X_1,\ldots,X_n$, the data consists of $(\tilde{X}_j,\delta_j) = (X_j \wedge X_{(r)}, \chi_{\{X_j \le X_{(r)}\}})$, $j = 1,\ldots,n$.

More generally, one might stop the experiment at some random "stopping time", based on the observed data at that moment. The data is now $(X_j \wedge T, \chi_{\{X_j \le T\}})$, $j = 1,\ldots,n$, where $T = T(X_1,\ldots,X_n)$ is such that $\chi_{\{T \le t\}}$ is some function of t and $(X_j \wedge t, \chi_{\{X_j \le t\}})$, $j = 1,\ldots,n$. RAO, SAVAGE & SOBEL (1960) give some examples of such censoring schemes in a two-sample situation.

This type of censorship is sometimes called "progressive censorship" but the term is more usually applied to the censorship discussed in Example 3.1.5.

EXAMPLE 3.1.3 "Random censorship", "competing risks".
In a biological experiment, one might observe the lifetimes of n experimental animals under certain conditions, together with the cause of death, which we suppose can be one of two types A or B. We are directly interested in the first of these two types - the animals may be divided into r groups according to different experimental conditions whose relation with A is to be investigated - while B comprises various accidental causes not directly related to the experiment. Let $\tilde{X}_j$ be the lifetime of the j-th animal, and let $\delta_j = 1$ or 0 according to whether it died from A or B. We suppose that different animals are independent of one another, and that given that animal j has survived up to time t, the conditional probability that it dies in the small time interval $[t,t+h]$ from cause A is approximately $\alpha_j(t)\cdot h$, while for B it is approximately $\beta_j(t)\cdot h$. Here α_j and β_j are continuous functions on $[0,\infty)$ called the forces of mortality for A and B; one would suppose that α_j is the same for experimental animals in the same group; β_j might be the same

for all animals, or it might vary from group to group or even within groups. In this situation $(\tilde{X}_j,\delta_j)$ can easily be shown to have the same distribution as $(X_j \wedge U_j, \chi_{\{X_j \leq U_j\}})$, where X_j and U_j are independent, with continuous densities $\alpha_j(t)\exp(-\int_0^t \alpha_j(s)ds)$ and $\beta_j(t)\exp(-\int_0^t \beta_j(s)ds)$. If for instance $\int_0^\infty \alpha_j(s)ds < \infty$, there is positive probability that $X_j = \infty$. Here, X_j can be thought of as the lifetime animal j would have had were β_j identically zero and thus cause B inoperative; while U_j is the conceptual lifetime of animal j were α_j identically zero.

So a model for this situation could consist of 2n independent positive- or infinite-valued random variables X_j, U_j; $j = 1,\dots,n$, from which the observed data $(\tilde{X}_j,\delta_j) = (X_j \wedge U_j, \chi_{\{X_j \leq U_j\}})$ is generated. X_j's within the same group will always be supposed to have the same distribution. Removing the implicit restriction to continuously distributed random variables, if the U_j's within the same group also have the same distribution this is known as "the model of random censorship". Our "general random censorship model" (see Example 4.1.1) will allow the U_j's to have arbitrary distributions.

Note that in general there is an identifiability problem; i.e. *dependent* X_j's and U_j's with different marginal distributions can lead to the same distribution for $(X_j \wedge U_j, \chi_{\{X_j \leq U_j\}})$ (see e.g. PETERSON (1975) and TSIATIS (1978)).

On the other hand one might even suppose that the U_j's are not independent of one another (e.g. animals, subject to an infectious disease, sharing a cage). However as long as $(X_1,\dots,X_n)$ is independent of $(U_1,\dots,U_n)$ this would not lead to problems.

EXAMPLE 3.1.4 "Fixed censorship", "progressive censorship of Type I". In a clinical trial, patients with a certain complaint entering a hospital between two fixed dates t_1 and t_2 are immediately given a treatment whose effectiveness is to be investigated at time t_2. Suppose that conditional on the number of patients $N = n$ entering between t_1 and t_2 and their entrance times $E_1 = e_1, \dots, E_n = e_n \in (t_1,t_2)$, the lengths of time $X_1,\dots,X_n$ elapsed between treatment time and time of eventual relapse are independent and identically distributed positive- or infinite-valued random variables. The aim is to say something about their common sub-distribution function F or to compare it with that associated with a different set of data pertaining to a different treatment. At time t_2 the available data is $(\tilde{X}_j,\delta_j) =$ $= (X_j \wedge u_j, \chi_{\{X_j \leq u_j\}})$, $j = 1,\dots,n$, where $u_j = t_2 - e_j$ is the fixed "observation limit" for the j-th patient (actually $u_1,\dots,u_n$ are also known and some statistical methods make use of them as well).

EXAMPLE 3.1.5 "Progressive censorship (of Type II)".
We return now to the industrial set-up described in Examples 3.1.1 and 3.1.3. Supposing the distribution of the n lifetimes $X_1,\dots,X_n$ to be continuous, the observation plan is now, at the time of the first observed failure time $X_{(1)}$, to remove from the test a random selection of r_1 components out of the still operating $n-1$. Supposing the $n-r_1-1$ remaining components to have lifetimes $Y_1,\dots,Y_{n-r_1-1}$, then at time $Y_{(1)}$, the next observed failure time, a further r_2 components are selected at random from those still on test and removed. This procedure is carried on till a total of s failures have been observed, with r_k components being withdrawn at the k-th stage, $k = 1,\dots,s$; $\sum_{k=1}^{s} (r_k+1) = n$. We now define $\tilde{X}_j = X_j$ and $\delta_j = 1$ if the j-th component is observed to fail at time X_j, and define $\tilde{X}_i = X_j$ and $\delta_i = 0$ if the i-th component is one of those removed at this time instant. The observed data is equivalent to $(\tilde{X}_j,\delta_j)$, $j = 1,\dots,n$. We say that component j is *on test* at time t if $\tilde{X}_j \geq t$, otherwise it has either failed or been removed at an earlier time instant.

Other terms such as "variable censorship" and "multiple censorship" occur in the literature, but generally one of the above examples is meant. All of these examples will be included in the general model of this section. Clearly various mixtures of these situations can also occur (and will also be included); for instance, in Example 3.1.4, the patients might also be subject to some "competing risks" such as accidental death from an unrelated cause, moving away from the district covered by a hospital, or whatever. Similarly in Example 3.1.3 there might be "planned withdrawals" of some of the surviving animals at fixed or random time instants for surgical investigations.

We next mention one example which will not be covered; we shall give it some attention in Chapter 6. The essential difference between this example and the previous ones is that the natural time axis in the new example does not permit one to consider each lifetime as starting on a new time axis at time $t = 0$, and still have cause and effect only working forwards in time. On the contrary, after this transformation the death or failure of one object at time t could effect the censoring of another at time $s < t$.

EXAMPLE 3.1.6 "Testing with replacement", "renewal testing".
Suppose that in Example 3.1.1, any component failing before time u is immediately replaced by a new one. So at any time instant up to u, exactly n components of varying age are on test. At the end of the test a random number of failures have been observed and there are exactly n censored observations.

We now state the model which will underlie the rest of this chapter and the following two chapters. Let (Ω,F,P) be a complete probability space on which are defined n independent positive, possibly infinite-valued random variables $X_1,\ldots,X_n$ with sub-distribution functions $F_1,\ldots,F_n$ defined by $F_j(t) = P(X_j \le t)$, $t \in [0,\infty)$, $F_j(\infty) = P(X_j < \infty)$. Define nondecreasing functions G_j with values in $\overline{\mathbb{R}}^+$ by

$$(3.1.1) \qquad G_j(t) = \int_{s\in[0,t]} (1 - F_j(s-))^{-1} dF_j(s).$$

Define

$$(3.1.2) \qquad \tau_j = \sup\{t: F_j(t) < 1\}.$$

We see that for each j, $F_j(0) = G_j(0) = 0$, G_j is finite on $[0,\tau_j)$, and G_j is constant on $[\tau_j,\infty]$. If $F_j(\tau_j-) < 1$ then G_j is bounded on $[0,\infty)$, and $\Delta G_j(\tau_j) = 1$ or 0 according to whether $\tau_j < \infty$ or $\tau_j = \infty$. In Lemma 3.2.1 we shall see that if on the other hand $F_j(\tau_j-) = 1$, then $G_j(t) \uparrow G_j(\tau_j) = \infty$ as $t \uparrow \tau_j$. If F_j has a density f_j, then defining the *hazard rate* $\lambda_j = f_j/(1-F_j)$ (in Example 3.1.3, $\lambda_j \equiv \alpha_j$), it holds for all t that $G_j(t) = \int_0^t \lambda_j(s)ds$. So G_j can be called the *cumulative hazard* or *cumulative risk* for the j-th object; see again Lemma 3.2.1.

We next suppose that $(\tilde{X}_j,\delta_j)$, $j = 1,\ldots,n$, are also defined on (Ω,F,P) and satisfy almost surely $0 < \tilde{X}_j < \infty$, $\tilde{X}_j \le X_j$, and $\delta_j = \chi_{\{\tilde{X}_j = X_j\}}$. Note that almost surely $G_j(\tilde{X}_j) \le G_j(X_j) < \infty$. We now define stochastic processes N_j, J_j and M_j, $j = 1,\ldots,n$, by

$$(3.1.3) \qquad N_j(t) = \chi_{\{\tilde{X}_j \le t, \delta_j = 1\}},$$

$$(3.1.4) \qquad J_j(t) = \chi_{\{\tilde{X}_j \ge t\}},$$

$$(3.1.5) \qquad M_j(t) = N_j(t) - G_j(\tilde{X}_j \wedge t) = N_j(t) - \int_0^t J_j dG_j.$$

We can now state our key model assumptions:

ASSUMPTION 3.1.1. There exist sub σ-algebras F_t of F making $(\Omega,F,P),\{F_t: t \in [0,\infty)\}$ a stochastic basis and N_j, J_j and M_j adapted processes for each j. M_j is a square integrable martingale for each j and $\langle M_j,M_j\rangle = \int J_j(1-\Delta G_j)dG_j$, $\langle M_j,M_{j'}\rangle = 0$ for all $j \neq j'$.

ASSUMPTION 3.1.2. For each $t \in [0,\infty)$, conditional on F_{t-}, $\Delta N_1(t),\ldots,\Delta N_n(t)$ are independent zero-one random variables with expectations $J_j(t)\Delta G_j(t)$, $j = 1,\ldots,n$.

We shall interpret these assumptions by relating them to the counting process theory of Section 2.3. It is convenient to consider the adaptedness requirements of Assumption 3.1.1 apart as a background assumption for both 3.1.1 and 3.1.2.

The adaptedness requirements are equivalent, given the stochastic basis $(\Omega,F,P),\{F_t: t \in [0,\infty)\}$, to requiring that $\chi_{\{\tilde{X}_j \le t\}}$, $\delta_j\chi_{\{\tilde{X}_j \le t\}}$ and $\tilde{X}_j\chi_{\{\tilde{X}_j \le t\}}$ are F_t-measurable for each t and j. In fact, Assumptions 3.1.1 and 3.1.2 are satisfied with respect to some stochastic basis if and only if they are satisfied with respect to the *minimal basis* defined by setting for each t

$$F_t = N \vee \sigma\{\chi_{\{\tilde{X}_j \le t\}}, \delta_j\chi_{\{\tilde{X}_j \le t\}}, \tilde{X}_j\chi_{\{\tilde{X}_j \le t\}},\ j = 1,\ldots,n\},$$

where N consists of all P-null sets of F and their complements. Whatever $\{F_t\}$ may be, we are supposing that the $\tilde{X}_j$'s are stopping times and that the events $\{\delta_j=0\}$ and $\{\delta_j=1\}$ happen at or before time $\tilde{X}_j$ (at time $\tilde{X}_j$, if $\{F_t\}$ is minimal). If the $\tilde{X}_j$'s are lifetimes, we are supposing that all lifetimes commence at time $t = 0$.

Given these background assumptions, Assumptions 3.1.1 and 3.1.2 in effect treat the continuous and the discrete cases respectively. If X_j has a continuous distribution for each j, Assumption 3.1.2 is empty; on the other hand, if X_j and $\tilde{X}_j$ are integer valued and $F_t = F_{[t]}$ for all $t \in [0,\infty)$, then Assumption 3.1.2 implies Assumption 3.1.1.

Now by the adaptedness requirements, N_j is a counting process and $\int J_j dG_j$ is predictable (for J_j is clearly predictable, and considered as a process, G_j is too). So requiring that M_j is a martingale is equivalent to requiring that N_j has compensator $\int J_j dG_j$. Thus $\int J_j dG_j$ can be thought of as the integrated conditional rate at which N_j jumps. We shall see presently that Assumptions 3.1.1 and 3.1.2 are satisfied if there is no censoring at all. So we are stating that at time t, given $\mathcal{F}_t$, if $\tilde{X}_j > t$ then N_j has the same conditional probability of jumping in the small time interval (t,t+h) as if there had been no censoring. As to what this rate is: if F_j has a continuous hazard rate λ_j, then this conditional probability is approximately $h\cdot\lambda_j(t)$. On the other hand, given $\mathcal{F}_t$, if $\tilde{X}_j \leq t$, then the conditional probability of jumping in (t,t+h) is zero.

The requirement that $\langle M_j,M_j\rangle = \int (1-\Delta G_j)J_j dG_j$ follows directly from Theorem 2.3.1 and need not have been made separately. If $F_1,\ldots,F_n$ are continuous then $\{N_1,\ldots,N_n\}$ forms a multivariate counting process and the requirement $\langle M_j,M_{j'}\rangle = 0$ also follows from Theorem 2.3.1. Otherwise it can be interpreted as a kind of pairwise independence condition, and it can in fact be derived from the following weaker version of Assumption 3.1.2: for each t and $j \neq j'$, conditional on $\mathcal{F}_{t-}$, $\Delta N_j(t)$ and $\Delta N_{j'}(t)$ are independent.

Assumption 3.1.2 itself is very simple to interpret, if we recall that $\Delta G_j(t) = P(X_j=t|X_j\geq t)$. Note also that $\tilde{X}_j\geq t \Rightarrow X_j\geq t$; and $\tilde{X}_j = t$ and $\delta_j=1 \Rightarrow X_j=t$. So we are stating that given what has happened up to but not including time t, if $\tilde{X}_j < t$, then the conditional probability that $\tilde{X}_j = t$ and $\delta_j = 1$ is zero; if $\tilde{X}_j \geq t$, then the probability that $\tilde{X}_j = t$ and $\delta_j = 1$ is equal to $P(X_j=t|X_j\geq t)$. Furthermore, still working conditionally on $\mathcal{F}_{t-}$, for j's such that $\tilde{X}_j \geq t$, the events $\{\tilde{X}_j=t,\delta_j=1\} = \{X_j=t\}$ are independent.

The next theorem gives an intuitively meaningful condition under which Assumptions 3.1.1 and 3.1.2 hold; as a corollary it follows that these assumptions hold in Examples 3.1.1 to 3.1.5 and when there is no censoring. The proofs of this and the following theorem simplify greatly when the F_j's are continuous.

<u>THEOREM 3.1.1</u>. *Let* $(\Omega,\mathcal{F},P),\{\mathcal{F}_t: t\in[0,\infty)\}$ *be a stochastic basis on which random variables* X_j, $\tilde{X}_j$ *and* δ_j $(j = 1,\ldots,n)$ *are defined, satisfying* $0 < \tilde{X}_j < \infty$, $\tilde{X}_j \leq X_j$ *and* $\delta_j = \chi_{\{\tilde{X}_j=X_j\}}$ *almost surely for each* j. *The* X_j*'s are supposed to be independent, with (sub)-distribution functions* F_j; *define* $G_j = \int (1-F_{j-})^{-1}dF_j$. *Suppose that* $\chi_{\{\tilde{X}_j\leq t\}}$ *and* $\delta_j\chi_{\{\tilde{X}_j\leq t\}}$ *are* $\mathcal{F}_t$*-measurable for each* j *and* t. *If for each* t, *conditional on* $\mathcal{F}_t$ *the* X_j*'s with* $\tilde{X}_j > t$ *are independent of one another, each having the distribution of* X_j *given* $X_j > t$, *then Assumptions* 3.1.1 *and* 3.1.2 *hold.*

<u>PROOF</u>. The measurability requirements of Assumption 3.1.1 follow directly from the measurability requirements of the theorem. Next, let I_1 and I_2 be disjoint sets of indices contained in $\{1,\ldots,n\}$ such that I_1 is nonempty; let j_0 be a fixed member of I_1; and define $I_0 = I_1\setminus\{j_0\}$. Consider the univariate counting process $N = \int \prod_{j\in I_0} \Delta N_j \cdot \prod_{j\in I_2}(1-\Delta N_j)\,dN_{j_0}$ which counts 1

at the single time instant t, if it exists, for which $\tilde{X}_j = t$ and $\delta_j = 1$ for all $j \in I_1$, provided that for no $j \in I_2$, $\tilde{X}_j = t$ and $\delta_j = 1$. (An empty product equals 1.) Fix $t < \infty$ such that $G_{j_0}(t) < \infty$ and define $t_{m,i} = i2^{-m}t$, $i = 0,\dots,2^m$; $m = 1,2,\dots$. For any m and $i < 2^m$, define the event $B_{m,i}$ by

$$B_{m,i} = \{\forall j \in I_1,\ \tilde{X}_j > t_{m,i} \text{ and } X_j \in (t_{m,i},t_{m,i+1}];\ \forall j \in I_2,\ \tilde{X}_j \le t_{m,i} \text{ or } (\tilde{X}_j > t_{m,i} \text{ and } X_j > t_{m,i+1})\}.$$

We shall approximate the increment of N over the interval $(t_{m,i},t_{m,i+1}]$ with $\chi_{B_{m,i}}$; in fact we have

$$(3.1.6)\qquad |(N(t_{m,i+1})-N(t_{m,i})) - \chi_{B_{m,i}}| \le \sum_{j\in I_1} \chi_{\{t_{m,i}<\tilde{X}_j<X_j\le t_{m,i+1}\}}$$
$$+ \sum_{j\neq j'\in I_1} \chi_{\{\tilde{X}_j,\tilde{X}_{j'}>t_{m,i};X_j,X_{j'}\in(t_{m,i},t_{m,i+1}];\ X_j\neq X_{j'}\}}$$
$$+ \sum_{j\in I_1,j'\in I_2} \chi_{\{\tilde{X}_j,\tilde{X}_{j'}>t_{m,i};X_j,X_{j'}\in(t_{m,i},t_{m,i+1}];\ X_j\neq X_{j'}\}}.$$

Now by the conditions of the theorem,

$$E(\chi_{B_{m,i}}|F_{t_{m,i}}) =$$
$$= \prod_{j\in I_1}\left(J_j(t_{m,i})\frac{F_j(t_{m,i+1}) - F_j(t_{m,i})}{1 - F_j(t_{m,i})}\right)\prod_{j\in I_2}\left(1 - J_j(t_{m,i})\frac{F_j(t_{m,i+1}) - F_j(t_{m,i})}{1 - F_j(t_{m,i})}\right) =$$
$$= \int_{s\in(t_{m,i},t_{m,i+1}]} \prod_{j\in I_0} J_j(t_{m,i})\frac{F_j(t_{m,i+1}) - F_j(t_{m,i})}{1 - F_j(t_{m,i})}\,\cdot$$
$$\cdot \prod_{j\in I_2}\left(1 - J_j(t_{m,i})\frac{F_j(t_{m,i+1}) - F_j(t_{m,i})}{1 - F_j(t_{m,i})}\right)\cdot J_{j_0}(t_{m,i})\,\frac{1}{1 - F_{j_0}(t_{m,i})}\,dF_{j_0}(s).$$

Thus

$$\sum_{i=0}^{2^m-1} E(\chi_{B_{m,i}}|F_{t_{m,i}}) = \int_0^t Y_m dF_{j_0},$$

where $0 \le Y_m(s) \le (1 - F_{j_0}(t-))^{-1} < \infty$ for all m and s and where

$$Y_m(s) \to \prod_{j\in I_0} J_j(s)\,\frac{\Delta F_j(s)}{1 - F_j(s-)}\prod_{j\in I_2}\left(1 - J_j(s)\frac{\Delta F_j(s)}{1 - F_j(s-)}\right)\frac{J_{j_0}(s)}{1 - F_{j_0}(s-)}$$

as $m \to \infty$ for all s, outside of an event of probability zero. Therefore, with the above choice of versions of $E(\chi_{B_{m,i}} | F_{t_{m,i}})$, we have

$$\sum_{i=0}^{2^m-1} E(\chi_{B_{m,i}} | F_{t_{m,i}}) \to \int_0^t \prod_{j\in I_0} J_j \Delta G_j \prod_{j\in I_2} (1 - J_j \Delta G_j) J_{j_0} dG_{j_0}$$

as $m \to \infty$ almost surely.

Next we consider the terms on the right hand side of (3.1.6). We have

$$0 \le E\Big(\sum_{i=0}^{2^m-1} E(\chi_{\{t_{m,i} < \tilde{X}_j < X_j \le t_{m,i+1}\}} | F_{t_{m,i}})\Big) \le P(0 < X_j - \tilde{X}_j < 2^{-m}t) \to 0$$

as $m \to \infty$. Similarly we can bound the expectation of the sum over i of conditional expectations of any of the other terms on the right hand side of (3.1.6) with $P(|X_j - X_{j'}| < 2^{-m}t,\ X_j \ne X_{j'}) \to 0$ as $m \to \infty$. Thus for any bounded random variable Y,

$$E\Big(Y \cdot \sum_{i=0}^{2^m-1} E(N(t_{m,i+1}) - N(t_{m,i}) | F_{t_{m,i}})\Big)$$

$$\to E\Big(Y \cdot \int_0^t \prod_{j\in I_0} J_j \Delta G_j \prod_{j\in I_2} (1 - J_j \Delta G_j) J_{j_0} dG_{j_0}\Big).$$

Let the compensator of N be A. By Theorem 2.3.2 we now have, for all $t < \infty$ such that $G_{j_0}(t) < \infty$,

$$A(t) = \int_0^t \prod_{j\in I_0} J_j \Delta G_j \prod_{j\in I_2} (1 - J_j \Delta G_j)\, J_{j_0} dG_{j_0} \quad \text{almost surely.} \tag{3.1.7}$$

We next show that A is constant on $[\tilde{X}_{j_0}, \infty)$. Define

$$T_\varepsilon = \inf\{t \ge \tilde{X}_{j_0} : A(t) - A(\tilde{X}_{j_0}) \ge \varepsilon\}, \qquad \varepsilon > 0,$$

where $\inf \emptyset = \infty$. $\tilde{X}_{j_0}$ and T_ε are stopping times, $\tilde{X}_{j_0} \le T_\varepsilon$, and by Theorem 2.3.1, $M = N - A$ is a martingale on $[0,\infty]$. So by Doob's optional stopping theorem,

$$E(N(T_\varepsilon) - N(\tilde{X}_{j_0})) = E(A(T_\varepsilon) - A(\tilde{X}_{j_0})) \ge \varepsilon\, P(T_\varepsilon < \infty).$$

But N is constant on $[\tilde{X}_{j_0}, \infty)$ so $P(T_\varepsilon < \infty) = 0$ for each $\varepsilon > 0$. With probability 1, $G_{j_0}(\tilde{X}_{j_0}) \le G_{j_0}(X_{j_0}) < \infty$. By right continuity of A, (3.1.7) with the fact that A is constant on $[\tilde{X}_{j_0}, \infty)$ shows that the processes A and

$$\int \prod_{j\in I_0} J_j\Delta G_j \prod_{j\in I_2} (1-J_j\Delta G_j)J_{j_0}\, dG_{j_0}$$

are indistinguishable.

Taking $I_1 = \{j\}$, $I_2 = \emptyset$ shows that N_j, defined by

$$N_j(t) = \chi_{\{\tilde{X}_j\le t,\delta_j=1\}}$$

has compensator $A_j = \int J_j dG_j$. Hence by Theorem 2.3.1, $\langle M_j,M_j\rangle =$ $= \int (1-J_j\Delta G_j)J_j dG_j$. To show that say $\langle M_1,M_2\rangle = 0$, consider the processes

$$N_1^* = \int (1-\Delta N_2)\,dN_1$$
$$N_2^* = \int (1-\Delta N_1)\,dN_2$$
$$N_3^* = \int \Delta N_1 dN_2.$$

Note that $\{N_1^*,N_2^*,N_3^*\}$ is a trivariate counting process, with compensators

$$A_1^* = \int (1-J_2\Delta G_2)J_1 dG_1$$
$$A_2^* = \int (1-J_1\Delta G_1)J_2 dG_2$$
$$A_3^* = \int J_1\Delta G_1 J_2 dG_2$$

by various choices of I_1 and I_2. Define $M_i^* = N_i^*-A_i^*$. Since $N_1^*+N_3^* = N_1$ and $N_2^*+N_3^* = N_2$ we also have $A_1^*+A_3^* = A_1$ and $A_2^*+A_3^* = A_2$. Therefore

$$\langle M_1,M_2\rangle = \langle M_1^*+M_3^*,M_2^*+M_3^*\rangle = \langle M_1^*,M_2^*\rangle + \langle M_1^*,M_3^*\rangle + \langle M_3^*,M_2^*\rangle + \langle M_3^*,M_3^*\rangle$$

$$= -\int \Delta A_1^* dA_2^* - \int \Delta A_1^* dA_3^* - \int \Delta A_3^* dA_2^* - \int \Delta A_3^* dA_3^* + A_3^*$$

(by Theorem 2.3.1)

$$= -\int \Delta(A_1^*+A_3^*)\,d(A_2^*+A_3^*) + A_3^*$$

$$= -\int \Delta A_1 dA_2 + \int \Delta A_1 dA_2$$

$$= 0.$$

This completes the proof that Assumption 3.1.1 holds. Now for any martingale M, $E(\Delta M(t)|F_{t-}) = 0$. Applied to the martingale $M = N-A$, we have $E(\Delta N(t)|F_{t-}) = \Delta A(t)$, i.e.

$$P(\Delta N_j(t) = 1 \ \forall j \in I_1,\ \Delta N_j(t) = 0 \ \forall j \in I_2 \mid F_{t-})$$

$$= \prod_{j \in I_1} J_j(t)\Delta G_j(t) \cdot \prod_{j \in I_2} (1 - J_j(t)\Delta G_j(t)),$$

which shows that Assumption 3.1.2 holds too. □

Considering the X_j's as lifetimes, commencing at time $t = 0$, we can interpret "$\tilde{X}_j > t$" as stating that the j-th object is under observation just after time t. So the intuitive content of Theorem 3.1.2 is that our assumptions hold if, for every t, given what has happened up to and including time t, the remaining lifetimes of the objects which are still under observation just after time t have the same joint distribution as if there had been no censoring. In particular, the fact that an object has not been censored in $[0,t]$ gives no information about its remaining life distribution. Such a condition is often used to give informal justification for various procedures in the analysis of censored data.

COROLLARY 3.1.1. *Assumptions* 3.1.1 *and* 3.1.2 *hold for Examples* 3.1.1 *to* 3.1.5.

PROOF. It is given that $X_1,\ldots,X_n$ are independent, with distribution functions $F_1,\ldots,F_n$. Examples 3.1.1, 3.1.3 and 3.1.4 are special cases of the following: $(U_1,\ldots,U_n)$ is independent of $(X_1,\ldots,X_n)$, and $\tilde{X}_j = X_j \wedge U_j$, $\delta_j = \chi_{\{X_j \le U_j\}}$ for each j. Example 3.1.2 is a special case of Example 3.1.5. In Example 3.1.5, suppose that the randomizations needed at the first $s-1$ stages in this example are generated by random vectors $V_1,\ldots,V_{s-1}$ (so V_k specifies which objects are to be removed from those remaining at stage k). Suppose that $X_1,\ldots,X_n$, $U_1,\ldots,U_n$ or $X_1,\ldots,X_n,V_1,\ldots,V_{s-1}$ are defined on a complete probability space (Ω,F,P); let N in each case be the σ-algebra of all P-null sets of F and their complements; and define

$$F_t = N \vee \sigma\{U_1,\ldots,U_n,\ \tilde{X}_j\chi_{\{X_j=\tilde{X}_j \le t\}}:\ j = 1,\ldots,n\}$$

or

$$F_t = N \vee \sigma\{V_1,\ldots,V_{s-1},\ \tilde{X}_j\chi_{\{X_j=\tilde{X}_j \le t\}}:\ j = 1,\ldots,n\}$$

for the first or second set of examples respectively. The conditions of Theorem 3.1.1 are now easy to verify (and the discussion in Appendix 2 shows that $(\Omega,F,P),\{F_t: t \in [0,\infty)\}$ is indeed a stochastic basis). □

Other choices of $\{\mathcal{F}_t\}$ in the proof of Corollary 3.1.1 would have been more natural and would also have satisfied the conditions of Theorem 3.1.1. However the above choice is useful in applying the next theorem to Examples 3.1.1 to 3.1.5. This theorem specifies the likelihood ratio based on the observations $(\tilde{X}_j,\delta_j)$, $j = 1,\ldots,n$ for the hypothesis H: X_j has distribution function F_j, $j = 1,\ldots,n$, and H': X_j has distribution function F_j', $j = 1,\ldots,n$. The conditions imply those of Theorem 3.1.1, both under H and H'; they are discussed after the proof. We shall only need this theorem in Chapter 5.

<u>THEOREM 3.1.2</u>. *Let* $(\Omega,\mathcal{F},P),\{\mathcal{F}_t: t \in [0,\infty)\}$ *and* $(\Omega,\mathcal{F},P'),\{\mathcal{F}_t: t \in [0,\infty)\}$ *form two stochastic bases, and let* X_j, $\tilde{X}_j$ *and* δ_j, $j = 1,\ldots,n$, *be random variables with the usual properties* $0 < \tilde{X}_j < \infty$, $\tilde{X}_j \le X_j$, $\delta_j = \chi_{\{X_j=\tilde{X}_j\}}$, $(j = 1,\ldots,n)$ *almost surely* P *and almost surely* P'; *suppose that* $X_1,\ldots,X_n$ *are independent under* P *and* P' *and that* $P(X_j\le t) = F_j(t)$, $P'(X_j\le t) = F_j'(t)$, $t \in [0,\infty)$, *for (sub)-distribution functions* F_j *and* F_j', $j = 1,\ldots,n$.

Suppose that under P *or* P', *for each* t, *conditional on* $\mathcal{F}_t$, *the* X_j*'s with* $\tilde{X}_j > t$ *are independent, each having the distribution of* X_j *given* $X_j > t$ *(corresponding to* P *or* P' *respectively). Suppose that*

$$\mathcal{F}_t = \mathcal{F}_0 \vee \sigma\{\tilde{X}_j\chi_{\{X_j=\tilde{X}_j\le t\}}: j = 1,\ldots,n\} \quad \textit{for all } t$$

$$\tilde{X}_j\chi_{\{\tilde{X}_j\le t\}} \textit{ is } \mathcal{F}_t\textit{-measurable for all } j \textit{ and } t,$$

P *and* P' *agree on* $\mathcal{F}_0$

and

P *and* P' *are absolutely continuous with respect to one another on* $\mathcal{F}_\infty$.

Then on $(\Omega,\mathcal{F}_\infty)$

$$(3.1.8)\qquad \frac{dP'}{dP} = \prod_{j:\delta_j=1} \frac{dF_j'}{dF_j}(\tilde{X}_j) \prod_{j:\delta_j=0} \frac{1 - F_j'(\tilde{X}_j)}{1 - F_j(\tilde{X}_j)}$$

$$= \prod_{j:\delta_j=1} \frac{1 - \Delta G_j(\tilde{X}_j)}{1 - \Delta G_j'(\tilde{X}_j)} \frac{dG_j'}{dG_j}(\tilde{X}_j) \cdot \prod_{j=1}^{n} \frac{1 - F_j'(\tilde{X}_j)}{1 - F_j(\tilde{X}_j)} .$$

PROOF. We apply Theorem 2.3.3 to the (2^n-1)-variate counting process with components indexed by the non-empty subsets of $\{1,\ldots,n\}$:

$$\left\{N_I = \int \prod_{j\in I\setminus\{j_0\}} \Delta N_j \prod_{j\notin I} (1-\Delta N_j)\,dN_{j_0} : I\subset\{1,\ldots,n\},\ I\neq\emptyset\right\},$$

where j_0 is an arbitrary member of I. As was seen in the proof of Theorem 3.1.1, N_I has (under P) compensator

$$A_I = \int \prod_{j\in I\setminus\{j_0\}} J_j\Delta G_j \prod_{j\notin I} (1 - J_j\Delta G_j)J_{j_0}\,dG_{j_0}.$$

Note also that the sum of all the components of the above counting process is the univariate counting process

$$\tilde{N} = \sum_{j=1}^{n} \int \prod_{j'<j} (1 - \Delta N_{j'})\,dN_j,$$

which counts 1 at each jump of $\sum_{j=1}^{n} N_j$, and which has compensator

$$\tilde{A} = \sum_{j=1}^{n} \int \prod_{j'<j} (1 - J_{j'}\Delta G_{j'})J_j\,dG_j.$$

We also have $\tilde{A}_c = \sum_{j=1}^{n} \int J_j\,dG_{jc}$ and $1 - \Delta\tilde{A} = \prod_{j=1}^{n} (1 - J_j\Delta G_j)$. Let $T_1 < \ldots < T_m$ be the distinct times at which $\tilde{N}$ jumps ($m = \tilde{N}(\infty)$ is random). By Theorem 2.3.3, on F_∞

$$\frac{dP'}{dP} = \prod_{\ell=1}^{m} \left(\prod_{j:\tilde{X}_j=T_\ell,\delta_j=1} \frac{dG'_j}{dG_j}(\tilde{X}_j)\cdot \prod_{\substack{j:\tilde{X}_j>T_\ell \text{ or}\\ \tilde{X}_j=T_\ell \text{ and } \delta_j=0}} \frac{1-\Delta G'_j(T_\ell)}{1-\Delta G_j(T_\ell)}\right)\cdot$$

$$\cdot\left(\prod_{s\notin\{T_1,\ldots,T_m\}} \frac{\prod_{j=1}^{n}(1-J_j(s)\Delta G'_j(s))}{\prod_{j=1}^{n}(1-J_j(s)\Delta G_j(s))}\right) \frac{\exp(-\sum_{j=1}^{n}\int_0^\infty J_j\,dG'_{jc})}{\exp(-\sum_{j=1}^{n}\int_0^\infty J_j\,dG_{jc})}$$

$$= \prod_{j:\delta_j=1} \frac{dG'_j}{dG_j}(\tilde{X}_j)\cdot \prod_{j:\delta_j=1} \frac{1-\Delta G_j(\tilde{X}_j)}{1-\Delta G'_j(\tilde{X}_j)}\cdot$$

$$\cdot \prod_{j=1}^{n} \frac{(\prod_s(1-J_j(s)\Delta G'_j(s)))\exp(-\int_0^\infty J_j\,dG'_{jc})}{(\prod_s(1-J_j(s)\Delta G_j(s)))\exp(-\int_0^\infty J_j\,dG_{jc})} =$$

$$= \prod_{j:\delta_j=1} \left(\frac{dG_j'}{dG_j}(\tilde{X}_j) \frac{1-\Delta G_j(\tilde{X}_j)}{1-\Delta G_j'(\tilde{X}_j)}\right) \cdot \prod_j \frac{1-F_j'(\tilde{X}_j)}{1-F_j(X_j)} \quad \text{by Lemma 3.2.1(i)}$$

$$= \prod_{j:\delta_j=1} \frac{dF_j'}{dF_j}(\tilde{X}_j) \cdot \prod_{j:\delta_j=0} \frac{1-F_j'(\tilde{X}_j)}{1-F_j(\tilde{X}_j)} . \qquad \square$$

The expression on the right hand side of (3.1.8) is often used as a likelihood ratio on intuitive grounds, see e.g. COX (1975) and BRESLOW (1975). Note that with the definition of $\mathcal{F}_t$ given in Corollary 3.1.1, the theorem applies to all of Examples 3.1.1 to 3.1.5, if changing P to P' only changes the distributions of the X_j's, and not of the U_j's or V_k's.

The extra condition in Theorem 3.1.2 on the σ-algebras $\mathcal{F}_t$ can be intuitively interpreted as requiring that all random aspects of the censoring, except in so far as they are generated by the lifetimes X_j themselves, can be conceived of as being realized at time $t = 0$, which is hardly a restriction at all. What is a restriction is that P and P' should agree on $\mathcal{F}_0$; i.e. censoring gives no information on which probability measure holds, except in so far as it depends on the X_j's.

3.2. One sample case: the product limit estimator

In this section we specialize the general model given after the examples of the previous section by supposing that $F_1 = \ldots = F_n = F$, say. Define $G = G_j$ (see 3.1.1), $\tau = \tau_j$ (3.1.2), and recall the definitions of N_j, J_j and M_j (3.1.3 to 3.1.5). We assume that Assumption 3.1.1 holds, but will not need Assumption 3.1.2 in this section.

The product limit estimator $\hat{F} = \{\hat{F}(t): t \in [0,\infty)\}$ is an estimator of F based on the observations $(\tilde{X}_j,\delta_j)$, $j = 1,\ldots,n$, which reduces to the usual empirical distribution function based on $X_1,\ldots,X_n$ if $\delta_j = 1$ for each j (recall that $\tilde{X}_j = X_j$ if $\delta_j = 1$, otherwise $\tilde{X}_j < X_j$ and $\delta_j = 0$, where the X_j's are independent and identically distributed with distribution function F). The estimator $\hat{F}$ was introduced in statistics by KAPLAN & MEIER (1958), and a closely related estimator of log(1-F) by NELSON (1972). However versions of it had long been known in the fields of demography and actuarial science. Recently, smoothed versions have been proposed ($\hat{F}$ itself is a step function), e.g. by AALEN & JOHANSEN (1978) and FÖLDES, REJTŐ & WINTER (1980). BARLOW & CAMPO (1975) propose another estimator of a certain transform of F, called the "total time on test plot". However there are some difficulties in applying this to censored data which have not been resolved yet. In Appendix 5 we make some suggestions in this direction.

$\hat{F}$ can be described as the sub-distribution function on $[0,\infty)$ which only assigns mass to the values of the uncensored observations, and which does this in such a way that for any $t \in [0,\infty)$,

$$\text{(3.2.1)} \qquad \frac{\Delta\hat{F}(t)}{1-\hat{F}(t-)} = \frac{\#\{j: \tilde{X}_j = t,\ \delta_j = 1\}}{\#\{j: \tilde{X}_j \geq t\}} .$$

When F is discrete, the right hand side of (3.2.1) is a very natural estimator of $P(X_j = t | X_j \geq t) = \Delta F(t)/(1-F(t-))$. $\hat{F}$ can often be thought of as the maximum likelihood estimator of F (the term needs qualification because in its usual sense, one does not exist, there being no dominating measure for the set of all measures on $[0,\infty)$, see e.g. JOHANSEN (1978)). It will be seen that the above definition allows $\hat{F}$ to be less than 1 and constant to the right of the largest observation $\tilde{X}_j$, if this observation or one of the group of tied largest observations is censored. Other definitions of the product limit estimator set it equal to 1 on this part of the real line, or leave it undefined there.

We presently give a concise definition of $\hat{F}$ in terms of the processes N_j and J_j, $j = 1,\ldots,n$, and establish some of its small sample properties. In Section 4.1 we prove consistency under a generalization of the random censorship model (covering Examples 3.1.1, 3.1.3 and 3.1.4) and in Section 4.2 we show how the estimator can be used to give confidence bands for the unknown F, and confidence intervals for F(t) for fixed t.

Define processes N, Y, M, J and the product limit estimator $\hat{F}$ by

$$\text{(3.2.2)} \qquad N(t) = \sum_{j=1}^{n} N_j(t) = \#\{j: \tilde{X}_j \leq t \text{ and } \delta_j = 1\}$$

$$\text{(3.2.3)} \qquad Y(t) = \sum_{j=1}^{n} J_j(t) = \#\{j: \tilde{X}_j \geq t\}$$

$$\text{(3.2.4)} \qquad M(t) = \sum_{j=1}^{n} M_j(t) = N(t) - \int_0^t Y\,dG$$

$$\text{(3.2.5)} \qquad J(t) = \chi_{\{Y(t) > 0\}}$$

and

$$\text{(3.2.6)} \qquad \hat{F}(t) = 1 - \prod_{s \leq t} \left(1 - \frac{\Delta N(s)}{Y(s)}\right)$$

where the convention 0/0 = 0 has been applied. N is nondecreasing and right continuous, Y is nonincreasing and left continuous; both take values in

$\{0,1,\ldots,n\}$. Also we have $Y(0) = n$ almost surely and $\Delta N(s) \le Y(s)$ for all s; if equality holds for some s then for $t > s$, $N(t) = N(s)$ and $Y(t) = 0$. In any case $Y(\infty) = 0$ almost surely. It is easy to check that (3.2.6) corresponds to the earlier verbal definition of $\hat{F}$. Since

$$(1-\hat{F}(t)) \prod_{s\le t}\left(1 - \frac{Y(s) - Y(s+) - \Delta N(s)}{Y(s) - \Delta N(s)}\right) = \frac{Y(t+)}{n},$$

we see that $(Y_+/n)/(1-\hat{F})$ is nonincreasing, nonnegative, and takes the value 1 at time zero (it can in fact be interpreted as 1 minus the product limit estimator of the censoring distribution). These facts give us in particular the right hand part of the inequality

$$N/n \le \hat{F} \le 1 - (Y_+/n).$$

The left hand part follows by comparing (3.2.6) with the equality

$$\frac{N(t)}{n} = 1 - \prod_{s\le t}\left(1 - \frac{\Delta N(s)}{n - N(s-)}\right).$$

Equivalent to (3.2.6) is the implicit definition

$$\hat{F}(t) = \int_{s\in[0,t]} (1-\hat{F}(s-))\,\frac{dN(s)}{Y(s)}\,. \tag{3.2.7}$$

Note that F and G satisfy

$$F(t) = \int_{s\in[0,t]} (1-F(s-))\,dG(s), \tag{3.2.8}$$

so it is not surprising that $\int Y^{-1}dN$, the so-called *empirical cumulative hazard function*, can be considered as an estimator of G; see e.g. NELSON (1972). The following lemma shows that given G, equation (3.2.8) implicitly determines F, which suggests why (3.2.7) and (3.2.8) will be so important: the closer $\int Y^{-1}dN$ is to G, the closer will $\hat{F}$ be to F. The proof is purely analytic and is given in Appendix 4.

<u>LEMMA 3.2.1</u>. *Let* $G = \int (1-F_-)^{-1}dF$ *for some (sub)-distribution function* F *with* $F(0) = 0$, *and define* $\tau = \sup\{t: F(t) < 1\}$.

(i) (3.2.8) *uniquely determines* F *if* G *is given; and* F *can be written as*

$$F(t) = 1 - \prod_{s\le t}(1 - \Delta G(s))\cdot\exp(-G_c(t)) \tag{3.2.9}$$

for all t.

(ii) *F and G are constant on $[\tau,\infty)$, G is finite and $\Delta G < 1$ on $[0,\tau)$. If $F(\tau-) < 1$, then $G(\tau) < \infty$ and $\Delta G(\tau) = 1$ iff $F(\tau) = 1$. If on the other hand $F(\tau-) = 1$, then $G(t) \uparrow G(\tau) = \infty$ as $t \uparrow \tau$.*

(iii) *If F has a density f, then defining the hazard rate or failure rate λ by $\lambda = f/(1-F)$,*

$$(3.2.10) \qquad G(t) = \int_{s\in[0,t]} \lambda(s)ds \quad \text{for all } t.$$

More generally, if F is only continuous, we have

$$(3.2.11) \qquad G = -\log(1-F).$$

(iv) *For all t such that $F(t) < 1$,*

$$(3.2.12) \qquad \frac{1-\hat{F}(t)}{1-F(t)} = 1 - \int_0^t \frac{1-\hat{F}(s-)}{1-F(s)}\left(\frac{dN(s)}{Y(s)} - dG(s)\right).$$

Relation (3.2.12) will later be extremely useful for deriving asymptotic results for $\hat{F}$. It can also be derived from Theorem 3.1 of AALEN & JOHANSEN (1978) who used it for the same purpose. In the meantime we shall couple (3.2.12) with Assumption 3.1.1 to derive some well-known results on $\hat{F}$.

Recalling the Definitions (3.2.4) and (3.2.5) of the processes J and M, and using (3.2.12), we see that for t such that $F(t) < 1$ and $Y(t) > 0$,

$$(3.2.13) \qquad \hat{F}(t) - F(t) = (1-F(t)) \int_0^t \frac{1-\hat{F}_-}{1-F}\frac{J}{Y}(dN - YdG)$$
$$= (1-F(t)) \int_0^t \frac{1-\hat{F}_-}{1-F}\frac{J}{Y}\, dM.$$

Let us define a stopping time T by

$$(3.2.14) \qquad T = \inf\{t: Y(t) = 0\}.$$

Note that $\hat{F}$ and M are constant on $[T,\infty)$ and that (3.2.13) holds with $t = T$ provided $F(T) < 1$. So for *any* t such that $F(t) < 1$,

$$(3.2.15) \qquad \hat{F}(t) - F(t) = (1-F(t)) \int_0^t \frac{1-\hat{F}_-}{1-F}\frac{J}{Y}\, dM$$
$$+ \chi_{\{T<t\}}\left(\hat{F}(t) - F(t) - (1-F(t)) \int_0^T \frac{1-\hat{F}_-}{1-F}\frac{J}{Y}\, dM\right) =$$

$$= (1-F(t)) \int_0^t \frac{1-\hat{F}_-}{1-F} \frac{J}{Y} dM + \chi_{\{T<t\}}\left(\hat{F}(T)-F(t) - (1-F(t))\frac{\hat{F}(T)-F(T)}{1-F(T)}\right)$$

$$= (1-F(t)) \int_0^t \frac{1-\hat{F}_-}{1-F} \frac{J}{Y} dM - \chi_{\{T<t\}}\frac{(1-\hat{F}(T))(F(t)-F(T))}{1-F(T)} .$$

Now by Assumption 3.1.1 and Definition (3.2.5), M is a square integrable martingale, while $\frac{1-\hat{F}_-}{1-F}\frac{J}{Y}$ is bounded on $[0,t]$ for each t with $F(t) < 1$ and is predictable (J, Y and $\hat{F}_-$ are left continuous adapted processes while F is a deterministic process). So by (2.2.1) we obtain on $\{t: F(t) < 1\}$

$$\text{(3.2.16)} \qquad E\hat{F} = F - E\left(\chi_{\{T<t\}}\frac{(1-\hat{F}(T))(F(t)-F(T))}{1-F(T)}\right).$$

So $\hat{F}$ is in general biased downwards, and is unbiased on $\{t: F(t) < 1\}$ if and only if almost surely, $\hat{F}(T) = 1$ or F is constant on $\{t: t \geq T$ and $F(t) < 1\}$. A sufficient condition for unbiasedness is that almost surely, $Y(\tau) > 0$ or for some $t < \tau$, $\Delta N(t) = Y(t)$; i.e. if the largest observation is less than τ, it, and all observations equal to it, must be uncensored. In this case, if $F(\tau) = 1$, then $\hat{F}(\tau) = 1$ almost surely and we have unbiasedness on $[0,\infty)$.

Relation (3.2.16) shows that the absolute value of the bias of $\hat{F}(t)$ increases as t increases, and yields the following bound (true for all t such that $F(t) < 1$):

$$\text{(3.2.17)} \qquad 0 \leq F(t) - E\hat{F}(t) \leq F(t)P(Y(t)=0).$$

This improves the result given as the theorem in Section 2.2 in MEIER (1975), which concerns a continuous distribution function F and the model of fixed censorship (Example 3.1.4), and gives a slightly weaker bound.

We next briefly study the variance of $\hat{F}-F$, corrected for its "random bias"; i.e. defining

$$\text{(3.2.18)} \qquad B = -\chi_{\{T<t\}}\frac{(1-\hat{F}(T))(F(t)-F(T))}{1-F(T)} ,$$

we look at the variance of

$$\hat{F}-F-B = (1-F)\int \frac{1-\hat{F}_-}{1-F}\frac{J}{Y} dM$$

(cf. (3.2.15)). We shall use (2.2.2). By Assumption 3.1.1 and Definition (3.2.4), $\langle M,M\rangle$ is given by

(3.2.19) $\langle M,M\rangle = \int Y(1-\Delta G)\,dG.$

So by (2.2.2), for $t < \tau$,

(3.2.20) $$\mathrm{var}(\hat{F}(t) - F(t) - B(t)) = E((\hat{F}(t) - F(t) - B(t))^2)$$

$$= (1-F(t))^2 \int_0^t E\left(\frac{(1-\hat{F}_-)^2 J}{Y}\right) \frac{1-\Delta G}{(1-F)^2}\,dG$$

$$= (1-F(t))^2 \int_0^t E\left(\frac{(1-\hat{F}_-)^2 J}{Y}\right) \frac{dF}{(1-F_-)^2(1-F)} \quad .$$

This suggests that the following quantity could be used as an estimate of the variance of $\hat{F}(t) - F(t)$ for asymptotic purposes:

(3.2.21) $$\hat{V}(t) = (1-\hat{F}(t))^2 \int_0^t \frac{J}{Y(1-\hat{F})}\,d\hat{F} = (1-\hat{F}(t))^2 \int_0^t \frac{dN}{Y(Y-\Delta N)} \, .$$

This is in fact the estimator proposed by KAPLAN & MEIER (1958), formula 2f; we investigate it further in Section 4.2. Using the inequality $Y/n \le 1 - \hat{F}_-$ and (A.4.7) it follows straightforwardly that

$$\hat{V}(t) \ge n^{-1}\hat{F}(t)(1-\hat{F}(t))$$

with equality if and only if there are no censored observations in $[0,t]$.

The next result gives an "in probability linear bound" for the product limit estimator. Similar results for the empirical distribution function are well known; see for instance the references in SHORACK & WELLNER (1978). In VAN ZUIJLEN (1978) Theorem 1.1 and Corollary 3.1 these results (still for the empirical distribution function) are generalized to the case of not necessarily identical or continuous distribution functions. We are still assuming that $F_1 = \ldots = F_n = F$, for some not necessarily continuous (sub)-distribution function F; and Assumption 3.1.1 is supposed to hold.

THEOREM 3.2.1. *Defining*

$$T = \sup\{t\colon Y(t) > 0\}$$

we have for all $\beta \in (0,1)$

(3.2.22) $$P(1-\hat{F} \le \beta^{-1}(1-F) \text{ on } [0,T]) \ge 1-\beta.$$

PROOF. Define

$$Z(t) = \frac{1-\hat{F}(t\wedge T)}{1-F(t\wedge T)}, \qquad t \in [0,\infty).$$

By (3.2.12), Z is a martingale on $[0,t]$ for every t such that $F(t) < 1$. So by Doob's submartingale inequality for every $\beta > 0$ we have

$$P(\sup_{s\in[0,t]} Z(s) \geq \beta^{-1}) \leq \beta\, E|Z(t)| = \beta\, E(Z(t)) = \beta\, E(Z(0)) = \beta.$$

So we have

$$P(1-\hat{F} \leq \beta^{-1}(1-F) \text{ on } [0,t\wedge T]) \geq 1-\beta.$$

Recalling that $\tau = \sup\{t: F(t) < 1\}$, by letting $t \uparrow \tau$ we find

$$P(1-\hat{F} \leq \beta^{-1}(1-F) \text{ on } [0,T]\setminus\{\tau\}) \geq 1-\beta.$$

If $F(\tau_-) = F(\tau)$, $P(\hat{F}(\tau) = \hat{F}(\tau_-)) = 1$. If $F(\tau_-) < F(\tau) = 1$, we have $P(T=\tau \text{ and } \hat{F}(\tau)=1) = P(T=\tau)$. So in both cases we obtain (3.2.22). □

The bound in (3.2.22) is surprisingly sharp; DANIELS (1945) and ROBBINS (1954) show that (3.2.22) holds with equality when there is no censoring and F is continuous. In Appendix 6 we present a proof inspired by TAKÁCS (1967) explaining why DANIELS' and ROBBINS' result is so simple and why in particular there is no dependence on n.

One might have expected (cf. VAN ZUIJLEN (1978) Theorem 1.1) that results similar to (3.2.22) on $P(1-\hat{F} \geq \beta(1-F) \text{ on } [0,T))$, $P(\hat{F} \leq \beta^{-1}F)$ and $P(\hat{F} \geq \beta F \text{ on } \{t: N(t) > 0, Y(t) > 0\})$, could be obtained for the product limit estimator. However we have not succeeded in deriving this kind of result in as much generality as in Theorem 3.2.1; fortunately we only need the following rather limited result in the sequel.

PROPOSITION 3.2.1. *Suppose that* $F_1 = \ldots = F_n = F$ *for some continuous (sub)-distribution function* F, *and suppose that Assumption* 3.1.1 *holds. Define*

$$\tilde{F} = (n\hat{F}+1)/(n+1).$$

Then for all $\varepsilon > 0$ *there exists* $\beta = \beta(\varepsilon) \in (0,1)$ *such that for any* $u \in [0,\infty)$ *and* $\alpha \in (0,1)$

$$P(\tilde{F} \geq \alpha\beta F \text{ on } [0,u]) \geq 1-\varepsilon-P(Y(u) \leq \alpha n). \tag{3.2.23}$$

<u>PROOF</u>. If $F(u) = 1$ then $Y(u) = 0$ almost surely, and (3.2.23) holds trivially for any $\varepsilon > 0$. So we let u and α be fixed, and suppose that $F(u) < 1$. Without loss of generality we may then also suppose that $G = \int (1-F)^{-1}dF$ is finite on $[0,\infty)$ and $G(\infty) = \infty$. Let $k = [\alpha n] + 1$. The events $\{Y(u) > \alpha n\}$ and $\{Y(u) \geq k\}$ are identical. Also by the inequality $\hat{F} \geq N/n$ (see the discussion after (3.2.6)) we have

$$\tilde{F} \geq (N+1)/(n+1) = \alpha(N+1)/(\alpha n+\alpha) \geq \alpha(N+1)/(k+1).$$

We shall establish (3.2.22) by constructing random variables $X_1^*,\ldots,X_k^*$ which are independent and identically distributed with distribution function F and satisfy, on the event $\{Y(u) \geq k\}$

$$N(t) \geq N^*(t) = \#\{i\colon X_i^* \leq t\} \quad \text{for all } t \in [0,u].$$

For then, by VAN ZUIJLEN (1977) Lemma 2.3.1 (or by the remarks preceding Theorem 1.4 in VAN ZUIJLEN (1978)),

$$P((N^*+1)/(k+1) \geq \beta F \text{ on } [0,\infty)) = 1 - \mathcal{O}(1) \quad \text{as } \beta \downarrow 0$$

uniformly in F and k, and (3.2.23) holds.

In fact only N^* will appear explicitly in the following construction. Let as usual N_i and J_i, $i = 1,\ldots,n$, be defined by

$$N_i(t) = \chi_{\{\tilde{X}_i \leq t \text{ and } \delta_i = 1\}}$$

$$J_i(t) = \chi_{\{\tilde{X}_i \geq t\}},$$

so that $N = \sum_{i=1}^n N_i$ and $Y = \sum_{i=1}^n J_i$. Extending $(\Omega,\mathcal{F},P),\{\mathcal{F}_t\colon t \in [0,\infty)\}$, define counting processes $N_{n+1},\ldots,N_{n+k}$ which are independent of the original sample space and of one another, and are such that each N_{n+i}, $i = 1,\ldots,k$, is a time inhomogeneous Poisson process with $\mathcal{E}(N_{n+i}(t)) = G(t)$ for all t. Under this extension

$$M_i = N_i - \int J_i dG, \quad i = 1,\ldots,n,$$

remain martingales, and

$$M_i = N_i - G, \quad i = n+1,\ldots,n+k$$

are martingales too. The idea of the proof is that $N_{n+1},\dots,N_{n+k}$ supply a reserve of processes jumping at the correct rate, so that by registering the jumps of some of the processes $N_1,\dots,N_{n+k}$ we obtain a new counting process which jumps at the same rate as N^*, defined by

$$N^*(t) = \#\{i = 1,\dots,k: X_i^* \le t\}.$$

We shall only need to draw on our reserve if less than k of $N_1,\dots,N_n$ are still available, i.e. if $Y < k$.

Let us define a process K as follows: $K(0) = k$, K is left continuous, nondecreasing, takes values in $\{k,k+1,\dots,n+k\}$, and only jumps at the times of the censored observations. It does this in such a way that if at time t, $J_i(t) = 1$, $J_i(t+) = 0$ and $\delta_i = 0$ for exactly r of the i's satisfying $i \le K(t)\wedge n$, then $K(t+) = K(t) + \ell$ where ℓ is the smallest positive integer such that exactly r of the i's between $K(t) + 1$ and $K(t) + \ell$ satisfy $i > n$ or $i \le n$ and $J_i(t+) = 1$. At time t we shall be registering the jumps of $N_1,\dots,N_{K(t)}$; so this definition ensures that if one of the N_i's whose jumps are being registered is censored, it is immediately replaced by a new one. Since there are at most n censored observations, K can never exceed the value 2n; we shall see presently that K actually does not exceed the value n+k so that we indeed only need to construct N_{n+i} for $i \le k$. Next we define processes J_i for $i = n+1,\dots,n+k$ by requiring these processes to be left continuous and $\{0,1\}$-valued and to satisfy $J_i(0) = 0$; J_i jumps to 1 at time t if and only if $K(t) < i$ but $K(t+) \ge i$; and J_i jumps back to zero at (i.e. just after) the first jump of N_i after t.

Finally we *define*

$$N^* = \sum_{i=1}^{n+k} \int \chi_{\{K\ge i\}} J_i dN_i .$$

Note the following facts. N^* is a counting process, because the N_i's with probability 1 never jump simultaneously. Because $M^*=\sum_{n=1}^{n+k} \int \chi_{\{K\ge i\}} J_i dM_i$ is a martingale, we find that the compensator of N^* is A^* defined by

$$A^*(t) = \sum_{i=1}^{n+k} \int_{s\in[0,t]} \chi_{\{K(s)\ge i\}} J_i(s)dG(s)$$

$$= \int_{s\in[0,t]} \left(\sum_{i=1}^{K(s)} J_i(s)\right)dG(s).$$

Now $\sum_{i=1}^{K} J_i = k - N_-^*$. For both members are left continuous and integer valued. Both take the value k at time zero. Finally, both have the same jumps at the same times: for each process only jumps when one of the J_i's jumps, and if at time t there are r_1 i's with $i \leq K(t)$, $J_i(t) = 1$, $J_i(t+) = 0$ and $\Delta N_i(t) = 0$ and r_2 i's with $i \leq K(t)$, $J_i(t) = 1$, $J_i(t+) = 0$ and $\Delta N_i(t) = 1$, then at time t+, K has increased to such a value that

$$\sum_{i=K(t)+1}^{K(t+)} J_i(t+) = r_1$$

while

$$\sum_{i=1}^{K(t)} J_i(t+) = \sum_{i=1}^{K(t)} J_i(t) - r_1 - r_2.$$

So $\sum_{i=1}^{K(t+)} J_i(t+) = \sum_{i=1}^{K(t)} J_i(t) - r_2$, while $N^*(t) = N^*(t-) + r_2$. From the fact $\sum_{i=1}^{K} J_i = k - N_-^*$ we deduce that $N^*(\infty) = k$. From this it follows that K indeed never exceeds n+k, for otherwise N^* would count a jump of N_{n+i}, $i \geq 1$, for more than k i's. From the relation $\sum_{i=1}^{K} J_i = k - N_-^*$ it also follows that $Y(u) \geq k$ implies $K(u) \leq n$. For suppose $Y(u) \geq k$ but $K(u) > n$. For some $t < u$ we would then have $K(t) \leq n$ and $K(t+) > n$, and $\sum_{i=1}^{K(t+)} J_i(t+) >$ $> Y(t+) \geq Y(u) \geq k$, implying that $N^*(t) < 0$.

We have now also shown that N^* has as compensator $\int (k - N_-^*)dG$. But by Corollary 3.1.1, N^* would also have this process as compensator were it defined by

$$N^*(t) = \#\{i: X_i^* \leq t\},$$

where $X_1^*,\ldots,X_k^*$ are independent and identically distributed with distribution function F. Hence by Theorem 2.3.4, N^* has the same probability distribution as if it were defined in this way. □

The restriction above to continuous distribution functions could have been dropped, but only at the cost of an even more complicated proof in which Assumption 3.1.2 would be needed. On the other hand, similar results to Proposition 3.2.1 can be obtained very easily from the results in VAN ZUIJLEN (1977,1978) under the general random censorship model (Example 4.1.1) by using the inequalities

$$N/n \leq \hat{F} \leq 1 - (Y_+/n)$$

and the fact that under this model, N/n and $1 - Y_+/n$ are empirical distribution functions of independent but not identically distributed random variables. With such an approach, no extra difficulties are involved if F is allowed to have jumps.

Finally we derive a minor result for later use:

PROPOSITION 3.2.2. *If* $F_1 = \ldots = F_n = F$ *and Assumption* 3.1.1 *holds, then*

$$\int (\Delta N-1)dN - \int Y(Y-1)\Delta G dG$$

is a zero mean martingale on the time interval $[0,\infty]$.

PROOF. First note that

$$\int (\Delta N-1)dN = -N + \int \Delta N dN = -N + \int (\Delta M+Y\Delta G)(dM+YdG)$$

$$= -N + \int \Delta M dM + \int Y^2\Delta G dG + \int Y\Delta M dG + \int Y\Delta G dM$$

$$= -N + \int \Delta M dM + \int Y^2\Delta G dG + 2\int Y\Delta G dM.$$

Now $\int Y\Delta G dM$ is a martingale on $[0,\infty]$, for $Y\Delta G$ is a bounded predictable process. By MEYER (1976) Theorem II.14, $\int \Delta M dM - \langle M,M\rangle$ is also a martingale on $[0,\infty]$. So in view of (3.2.19)

$$\int (\Delta N-1)dN - \int Y(Y-1)\Delta G dG$$

$$= \int (\Delta N-1)dN + \int YdG - \int Y(1-\Delta G)dG - \int Y^2\Delta G dG$$

is a martingale on $[0,\infty]$, zero at time zero. □

3.3. Two sample case: the test statistics of Gehan, Efron and Cox

We now introduce, as members of a whole class of test statistics the three test statistics whose study will take up a major part of this work. All are nonparametric in the sense that few assumptions have to be made in order that they can be used to construct an approximate (i.e. asymptotically valid) test for the null-hypothesis of interest; however only in special cases can they be used to give a truly nonparametric test, in the sense that their null-hypothesis distribution is known. We discuss this point further after the necessary notation has been introduced.

Again we specialize the model given after the examples in Section 3.1, this time supposing that the n observations fall into two groups, in each of which the distribution functions F_j are the same. Relabeling the observations, we now suppose that the available data consists of $(\tilde{X}_{ij},\delta_{ij})$, $j = 1,\ldots,n_i$; $i = 1,2$; where the distribution function F_{ij} belonging to observation (i,j) satisfies $F_{ij} = F_i$ for each i and j. In Definitions (3.1.1) to (3.1.5) we replace the index j everywhere with (i,j), and define $G_i = G_{ij}$ and $\tau_i = \tau_{ij}$. Assumption 3.1.1 is again supposed to hold, and the null-hypothesis H_0 we want to test is that $F_1 = F_2$.

Next we define for each of the two samples $i = 1$ and $i = 2$ processes N_i, Y_i, M_i, J_i and $\hat{F}_i$ similarly to (3.2.2) to (3.2.6):

(3.3.1) $$N_i(t) = \sum_{j=1}^{n_i} N_{ij}(t) = \#\{j: \tilde{X}_{ij} \le t \text{ and } \delta_{ij} = 1\}$$

(3.3.2) $$Y_i(t) = \sum_{j=1}^{n_i} J_{ij}(t) = \#\{j: \tilde{X}_{ij} \ge t\}$$

(3.3.3) $$M_i(t) = \sum_{j=1}^{n_i} M_{ij}(t) = N_i(t) - \int_0^t Y_i(s)dG_i(s)$$

(3.3.4) $$J_i(t) = \chi_{\{Y_i(t)>0\}}$$

(3.3.5) $$\hat{F}_i(t) = 1 - \prod_{s\le t}\left(1 - \frac{\Delta N_i(s)}{Y_i(s)}\right).$$

$\hat{F}_i$ is now the product limit estimator for sample i.

By Assumption 3.1.1, M_1 and M_2 are square integrable zero mean martingales, with

(3.3.6) $$\langle M_i,M_i\rangle = \int Y_i(1-\Delta G_i)dG_i \qquad i = 1 \text{ or } 2$$

and

(3.3.7) $$\langle M_1,M_2\rangle = 0;$$

Y_1, Y_2, J_1 and J_2 are predictable processes.

In motivating a certain class of test statistics we shall begin by supposing that the alternative hypothesis of interest is H_1:

$$\text{"}\frac{dG_1}{d\mu}(t) \ge \frac{dG_2}{d\mu}(t) \text{ for } \mu\text{-almost all } t \in [0,\tau_1\wedge\tau_2]\text{"}$$

where μ is any σ-finite measure on $[0,\infty)$ dominating both G_1 and G_2 (e.g.

the sum of the measures generated by G_1 and G_2). So if F_1 and F_2 have densities with respect to Lebesgue measure, and hence the hazard rates λ_1 and λ_2 exist, the alternative hypothesis reduces to

$$"\lambda_1(t) \geq \lambda_2(t) \text{ for Lebesgue-almost all } t \in [0,\tau_1\wedge\tau_2]";$$

while if F_1 and F_2 each assign mass 1 to the positive integers, it reduces to

$$"P(X_1=t|X_1\geq t) \geq P(X_2=t|X_2\geq t) \text{ for each } t \in \mathbb{N} \cap [1,\tau_1\wedge\tau_2]"$$

(here X_1 and X_2 are random variables with distribution functions F_1 and F_2). We call H_1 the alternative of *ordered hazards*. By (3.2.9), if H_1 is true then for all t, $F_1(t) \geq F_2(t)$; i.e. we have a strong form of the commonly considered alternative of *stochastic ordering*.

Let K be a bounded nonnegative predictable process which is a function of the observations and which satisfies $Y_1(t) \wedge Y_2(t) = 0 \Rightarrow K(t) = 0$; we denote by $\mathcal{K}^+$ the class of all such processes. (The class $\mathcal{K}$ is defined in the same way, dropping the requirement that K be nonnegative.) We shall use $K \in \mathcal{K}^+$ as a random weight function with which estimates of $dG_1 - dG_2$, i.e. $\frac{dN_1}{Y_1} - \frac{dN_2}{Y_2}$, are combined for those t for which estimation is possible, i.e. for which $Y_1(t)$ and $Y_2(t)$ are positive. For given $K \in \mathcal{K}$ or $\mathcal{K}^+$, define

$$(3.3.8) \qquad W = \int K\frac{dN_1}{Y_1} - \int K\frac{dN_2}{Y_2}$$

and

$$(3.3.9) \qquad Z = \int \frac{K}{Y_1}dM_1 - \int \frac{K}{Y_2}dM_2 = W - \int K(dG_1 - dG_2) \quad \text{by (3.3.3).}$$

We now see by (2.2.1) that $EZ = 0$ so that under H_0, $EW(\infty) = 0$, while under H_1, if $K \in \mathcal{K}^+$, $EW(\infty) \geq 0$. Also, by the assumptions on K, $W(\infty)$ is an observable quantity. It seems reasonable to investigate whether a test of H_0 versus H_1 can be based on $W(\infty)$.

There are now two possibilities. Sometimes, a test can be carried out using a permutation distribution of $W(\infty)$ under H_0. This would for instance be the case (for sensible choices of K) in Example 3.1.3 if under H_0 the forces of mortality for the competing risks are identical for all animals, or in Example 3.1.4 if the two samples arise by assigning one of two treatments at random to each patient entering. However, unless the data comes from a well planned experiment, only rarely will this approach be possible.

Alternatively, and this will be our approach, one could rely on large sample results and suppose that under H_0, $W(\infty)$ is approximately $N(0,\sigma^2)$ distributed for some σ^2 which will have to be estimated.

In view of (2.2.2), (3.3.6) and (3.3.7) we find that

$$\mathcal{E}Z^2 = \mathcal{E}\left(\sum_{i=1}^{2} \int \frac{K^2}{Y_i}(1 - \Delta G_i)dG_i\right), \tag{3.3.10}$$

where under H_0, $Z = W$.

Recalling that $\int \frac{dN_i}{Y_i}$ can be considered as an estimator of G_i (and under H_0, $\int \frac{d(N_1+N_2)}{Y_1+Y_2}$ as an estimator of $G_1 = G_2$), we propose as alternative estimators for σ^2, $V_1(\infty)$ and $V_2(\infty)$, where V_1 and V_2 are defined by

$$V_1 = \sum_{i=1}^{2} \int \frac{K^2}{Y_i}\left(1 - \frac{\Delta N_i - 1}{Y_i - 1}\right)\frac{dN_i}{Y_i} \tag{3.3.11}$$

and

$$V_2 = \int \sum_{i=1}^{2} \frac{K^2}{Y_i}\left(1 - \frac{\Delta N_1 + \Delta N_2 - 1}{Y_1 + Y_2 - 1}\right)\frac{d(N_1+N_2)}{Y_1 + Y_2}\ . \tag{3.3.12}$$

More explicitly, the suggested test procedure is to reject H_0 in favour of H_1, if $W(\infty)V_1(\infty)^{-\frac{1}{2}}$ (or alternatively $W(\infty)V_2(\infty)^{-\frac{1}{2}}$) takes on too large a value as compared with the standard normal distribution. By an abuse of notation, we shall say that $W(\infty)V_\ell(\infty)^{-\frac{1}{2}}$, $\ell = 1$ or 2, is a *test statistic of the class* $\mathcal{K}$ or $\mathcal{K}^+$ according to whether $K \in \mathcal{K}$ or $K \in \mathcal{K}^+$. If $K \in \mathcal{K}^+$ and T is a stopping time depending on the observations, then $K\chi_{[0,T]} \in \mathcal{K}^+$ too. So for any such stopping time, $W(T)V_\ell(T)^{-\frac{1}{2}}$ is also a test statistic of the class $\mathcal{K}^+$. In particular we can take $T = t$ for any fixed $t \in [0,\infty]$. Similar statements hold for $K \in \mathcal{K}$.

The -1's in numerator and denominator of the terms in (3.3.11) and (3.3.12) standing for ΔG_i in (3.3.10) have been introduced for two reasons. In the first place, if F_1 and F_2 are continuous these terms with probability 1 disappear, making V_1 and V_2 simpler to calculate and also, as we shall see presently, correspond more closely to the relevant quantities for the test statistics of interest as they were originally proposed. Secondly, they make $V_2(\infty)$, and in some cases $V_1(\infty)$ too, an unbiased estimator of the null hypothesis variance of $W(\infty)$, as the following proposition shows.

PROPOSITION 3.3.1. *Under* H_0, $EV_2(\infty) = \mathrm{var}(W(\infty))$. *If* $Y_1(t) \wedge Y_2(t) \le 1 \Rightarrow$ $\Rightarrow K(t) = 0$, *then* $EV_1(\infty) = \mathrm{var}(Z(\infty))$ $(= \mathrm{var}(W(\infty))$ *under* $H_0)$.

PROOF. By Proposition 3.2.2 and (2.2.1) applied to the martingale $\int (\Delta N_i - 1)dN_i - \int Y_i(Y_i-1)\Delta G_i dG_i$ and to the bounded predictable process $K^2(Y_i^2(Y_i-1))^{-1}$,

$$EV_1 = \sum_{i=1}^{2} \int E\left(\frac{K^2}{Y_i^2} Y_i - \frac{K^2}{Y_i(Y_i-1)} Y_i(Y_i-1)\Delta G_i\right) dG_i$$

$$= \sum_{i=1}^{2} \int E\left(\frac{K^2}{Y_i}(1 - \Delta G_i)dG_i\right) \qquad \text{if } Y_i(t)(Y_i(t)-1) = 0 \Rightarrow K(t) = 0$$

$$= EZ^2 \qquad \text{by (3.3.10).}$$

This proves the statements on V_1. For V_2, we proceed similarly, applying Proposition 3.2.2 with $N = N_1 + N_2$, $Y = Y_1 + Y_2$ and $G = G_1 = G_2$. However since $(Y_1(t) = 0$ or $Y_2(t) = 0) \Rightarrow K(t) = 0$, it now follows that $Y(t)(Y(t) - 1) = 0 \Rightarrow K(t) = 0$, so no additional condition has to be made. □

We now show that subject to some minor modifications, the test statistics of GEHAN (1965), EFRON (1968) and COX (1972) are members of the class K^+. Define as in AALEN (1978)

(3.3.14) $$K_G = Y_1 Y_2$$

(3.3.15) $$K_E = (1-\hat{F}_{1-})(1-\hat{F}_{2-})J_1 J_2$$

(3.3.16) $$K_C = \frac{Y_1 Y_2}{Y_1 + Y_2}$$

and the associated processes W_G, Z_G, V_{1G}, V_{2G}, etc. (see (3.3.8), (3.3.9), (3.3.11) and (3.3.12)). Note that each of these K's is predictable, bounded and nonnegative, and depends only on the observations $(\tilde{X}_{ij}, \delta_{ij})$, $j = 1,\ldots,n_i$; $i = 1,2$. Then we find that $W_G(\infty)$ is the test statistic of GEHAN (1965) defined below his formula (3.1) if we let his x_i's correspond to our second sample and his y_j's correspond to our first sample. GEHAN (1965) bases a permutation test on $W_G(\infty)$ in the following way. Let $N_1 + N_2 = N$ and $Y_1 + Y_2 = Y$ and let $T_1 < \ldots < T_r$ be the different time instants at which N jumps (so r is a random variable too). Put $T_0 = 0$ and $T_{r+1} = \infty$.

GEHAN calls the collection

$$(3.3.17)\qquad \mathcal{P} = \{r;\ (\Delta N(T_i), Y(T_i)-Y(T_{i+1}) - \Delta N(T_i)),\quad i = 0,\ldots,r\}$$

the *pattern* of the combined sample. Here, $\Delta N(T_i)$ is the number of uncensored observations at T_i, while $Y(T_i) - Y(T_{i+1}) - \Delta N(T_i)$ is the number of censored observations falling in the interval $[T_i, T_{i+1})$. GEHAN now supposes that under H_0 and conditional on $\mathcal{P}$, the joint distribution of the $2(r+1)$ numbers of observations from the first sample in each of these categories is the same as that obtained by selecting at random n_1 objects out of a total of n_1+n_2, which are distributed over $2(r+1)$ cells according to the numbers in $\mathcal{P}$. For small samples the test can be based on the exact permutation distribution of $W_G(\infty)$ conditional on $\mathcal{P}$. However for larger samples GEHAN proposes a normal approximation based on the exact permutation expectation and variance of $W_G(\infty)$; he shows that under the permutation hypothesis

$$(3.3.18)\qquad E(W_G(\infty)\,|\,\mathcal{P}) = 0$$

and also calculates $\mathrm{var}(W_G(\infty)\,|\,\mathcal{P})$; we give it in a simpler form due to MANTEL (1967), which we also rewrite in a form more suited to our notation:

$$(3.3.19)\qquad \mathrm{var}(W_G(\infty)\,|\,\mathcal{P}) = \frac{n_1n_2}{(n_1+n_2)(n_1+n_2-1)}\left(\int_0^\infty (Y-N)^2 dN + \int_0^\infty N^2 d(Y(0) - Y_+ - N)\right)$$

$$= \frac{n_1n_2}{(n_1+n_2)(n_1+n_2-1)}\int_0^\infty Y(Y-\Delta N)\,dN.$$

GEHAN's proof that, in a special case of Example 3.1.4, conditional on $\mathcal{P}$ and under H_0, $W_G(\infty)/\sqrt{\mathrm{var}(W_G(\infty)\,|\,\mathcal{P})}$ is asymptotically standard normally distributed, and his proof of consistency of the corresponding test versus alternatives of stochastic ordering, require that F_1 and F_2 give mass 1 to a finite set of points. However a more generally applicable proof can be based on a theorem of WALD, WOLFOWITZ, NOETHER & HOEFFDING given in PURI & SEN (1971) page 73, together with MANTEL's (1967) representation of $W_G(\infty)$ as a "linear permutation test statistic"; see BETHLEHEM, DOES & GILL (1977).

BRESLOW (1970) considers $W_G(\infty)$ from a purely "large-sample" point of view under the random censorship model (Example 3.1.4); i.e. without assuming that under the null-hypothesis a permutation distribution is availabe. He suggests estimating the null-hypothesis variance of $W_G(\infty)$ with

$$(3.3.20)\qquad \int_0^\infty Y_1Y_2\,d(N_1+N_2) + \frac{1}{n_1}\int_0^\infty Y_1(n_1-Y_1)\,dN_2 + \frac{1}{n_2}\int_0^\infty Y_2(n_2-Y_2)\,dN_1 .$$

He assumes continuous F_1 and F_2; in which case the first term of the above estimator is almost surely equal to $V_{2G}(\infty)$. The other two terms will generally be asymptotically negligeable compared to the first.

EFRON (1967) proposed a test statistic $\hat{W}$ and sketched its large-sample properties under the condition that there be no ties between the $\tilde{X}_{ij}$'s; he too worked under the random censorship model. Letting his x_i's correspond to our first sample, and his y_j's to our second sample, $\hat{W}$ is defined by

$$(3.3.21)\qquad \hat{W} = -\int_{s\in(0,\infty)} (1-\hat{F}_1(s-))J_1(s)\,d((1-\hat{F}_2(s))J_2(s+)) .$$

$\hat{W}$ can be considered as an estimator of $P(X_1 \geq X_2)$, where X_1 and X_2 are independent random variables with distribution functions F_1 and F_2. So under H_0, $\hat{W}$ should approximately equal $\frac{1}{2}$.

Letting $T_i = \max_j \tilde{X}_{ij}$ and $T = T_1 \wedge T_2$, we see that

$$(3.3.22)\qquad \hat{W} = -\int_{s\in(0,\infty)} (1-\hat{F}_1(s-))J_1(s)\,d(1-\hat{F}_2(s))$$

$$+ \chi_{\{T_2\leq T_1\}}(1-\hat{F}_1(T-))(1-\hat{F}_2(T-))$$

$$= \int_0^\infty (1-\hat{F}_{1-})(1-\hat{F}_{2-})J_1J_2\frac{dN_2}{Y_2} + \chi_{\{T_2\leq T_1\}}(1-\hat{F}_1(T-))(1-\hat{F}_2(T-))$$

by (3.2.7).

By integrating (3.3.21) by parts, and supposing there to be no ties among the $\tilde{X}_{ij}$'s, we also find that

$$(3.3.23)\qquad \hat{W} = 1 + \int_{s\in(0,\infty)} (1-\hat{F}_2(s-))J_2(s)\,d((1-\hat{F}_1(s))J_1(s+))$$

and hence repeating the previous calculations and adding, we find

$$(3.3.24)\qquad 2\hat{W}-1 = -W_E(\infty) + (\chi_{\{T_2<T_1\}} - \chi_{\{T_1<T_2\}})(1-\hat{F}_1(T-))(1-\hat{F}_2(T-)).$$

The last term here will be negligeable compared to the first one under the conditions EFRON (1967) envisaged for his asymptotic results. However if (3.3.24) is used to extend the definition of $\hat{W}$ to tied $\tilde{X}_{ij}$'s, even if F_1 and F_2 are continuous (as in Example 3.1.1) this last term can cause disastrous behaviour of $\hat{W}$ so it seems better to redefine $\hat{W}$ as $\frac{1}{2} - \frac{1}{2}W_E(\infty)$; we shall only consider $W_E(\infty)$ in the sequel.

As an estimator of the asymptotic null-hypothesis variance of $2\hat{W} - 1$, EFRON (1967, formula 8.12 and later remarks) proposed the estimator (modulo end effects similar to those in (3.3.24))

$$\int_0^\infty \frac{(1-\hat{F}_{1-})^3}{Y_1} d\hat{F}_1 + \int_0^\infty \frac{(1-\hat{F}_{2-})^3}{Y_2} dF_2$$

$$= \int_0^\infty (1-\hat{F}_{1-})^4 \frac{dN_1}{Y_1^2} + \int_0^\infty (1-\hat{F}_{2-})^4 \frac{dN_2}{Y_2^2},$$

where the second form suggests that this estimator will be close to $V_{1E}(\infty)$ under the null hypothesis (when $\hat{F}_1$ and $\hat{F}_2$ will be close to one another) if F_1 and F_2 are continuous. In the sequel we will however only consider V_1 and V_2. Both the test statistics of GEHAN and EFRON simplify to the Wilcoxon test when there is no censoring.

Finally we consider $W_C(\infty)$. COX (1972) considers treating a certain statistic $U(0)/\sqrt{I(0)}$ as approximately standard normally distributed for generating a two-sided test of H_0 versus

$$H_1': \text{"}(1-\Delta G_1)^{-1} \frac{dG_1}{d\mu} = c(1-\Delta G_2)^{-1} \frac{dG_2}{d\mu} \text{ for some } c \neq 1\text{"}$$

where μ, supposed to dominate G_1 and G_2, is either Lebesgue measure or counting measure. (In the first case $\Delta G_i = 0$ and we speak of a "proportional hazards model"; in the second we have a "proportional odds model".) It turns out that calling COX's sample 0 and sample 1 our sample 2 and sample 1 respectively,

$$U(0) = W_C(\infty)$$

$$I(0) = V_{2C}(\infty).$$

In various special cases, THOMAS (1969 and 1975), CROWLEY & THOMAS (1975) and AALEN (1976) show that under H_0, $U(0)/\sqrt{I(0)}$ has asymptotically a standard normal distribution.

Other authors, e.g. KALBFLEISCH & PRENTICE (1973) and BRESLOW (1974) propose slight variations of $I(0)$ for the case when ties are present. However these are either proposals for dealing with originally continuous data which later has been grouped (as in MANTEL's (1967) and BRESLOW's

(1970) discussion of the effect of ties on GEHAN's (1965) test statistic), or the authors have other alternative hypotheses in mind.

The test statistic of COX has also been derived by MANTEL (1966), PETO (1972), PETO & PETO (1972) and THOMAS (1969) and is widely known as the *log rank test* and as the (generalized) Savage test. If F_1 and F_2 are continuous and H_1' holds for an arbitrary σ-finite measure dominating both G_1 and G_2, then by (3.2.17), $(1-F_1) = (1-F_2)^c$, a so-called Lehmann alternative (SAVAGE (1956)).

CHAPTER 4

ASYMPTOTIC RESULTS

4.1. Consistency of the product limit estimator and of test statistics of the class K^+

In this section we apply the theorem of LENGLART (Theorem 2.4.2 above) to obtain conditions for uniform consistency of the product limit estimator. We also use it, in a two sample situation, to obtain conditions under the alternative hypothesis for a test statistic of the class K^+ to converge in probability to infinity as the sample sizes tend to infinity. Since in Section 4.3 we show that such a test statistic is asymptotically normally distributed under the null hypothesis, this constitutes a demonstration of consistency against the alternatives considered. The restriction from the class K to the class K^+ is related to our choice of alternative hypotheses, all of which state in some sense that the observations in one sample are smaller than those in the other. We specialize the results to a general random censorship model (Example 4.1.1 below) and, as far as the test statistics are concerned, to those of GEHAN, EFRON and COX.

First of all we collect the most important definitions and assumptions used throughout Chapters 4 and 5. We suppose that for each n = 1,2,... the model for n censored observations specified after the examples in Section 3.1 is given. In particular, we shall make continued use of Assumption 3.1.1 and, after this section, of Assumption 3.1.2 also. The underlying probability space (and hence also the distribution functions concerned) may be different for each n. We indicate dependence on n (of a distribution function, for instance) by a superscript; however in most other cases this dependence is suppressed in our notation (in particular, as far as stochastic processes defined for each n are concerned). We introduce the notation for an r-sample set-up. In future only the cases r = 1 and r = 2 will be considered, and dealing with the case r = 1 we shall drop the index i = 1,...,r altogether.

So r is fixed and for each n = 1,2,... a stochastic basis is given on which random variables X_{ij}^n, $\tilde{X}_{ij}^n$ and δ_{ij}^n are defined, $j = 1,\dots,n_i$, $i = 1,\dots,r$, where the number of observations in the i-th sample $n_i = n_i(n)$ satisfies $\sum_{i=1}^r n_i = n$. We suppose that the X_{ij}^n's are independent, X_{ij}^n having (sub)-distribution function F_i^n, and $\tilde{X}_{ij}^n$ and δ_{ij}^n satisfying $0 < \tilde{X}_{ij}^n < \infty$, $\tilde{X}_{ij}^n \le X_{ij}^n$ and $\delta_{ij}^n = \chi_{\{\tilde{X}_{ij}^n = X_{ij}^n\}}$ almost surely. For i = 1,...,r and for each n we define stochastic processes by

$$(4.1.1) \qquad N_i(t) = \#\{j: \tilde{X}_{ij}^n \le t \text{ and } \delta_{ij}^n = 1\}$$

$$(4.1.2) \qquad Y_i(t) = \#\{j: \tilde{X}_{ij}^n \ge t\}$$

$$(4.1.3) \qquad M_i(t) = N_i(t) - \int_0^t Y_i(s)\,dG_i^n(s)$$

$$(4.1.4) \qquad J_i(t) = \chi_{\{Y_i(t)>0\}}$$

$$(4.1.5) \qquad \hat{F}_i(t) = 1 - \prod_{s\le t}\left(1 - \frac{\Delta N_i(s)}{Y_i(s)}\right).$$

The function G_i^n in (4.1.3) is defined by $G_i^n = \int (1 - F_{i-}^n)^{-1} dF_i^n$. We also define $\tau_i^n = \sup\{t: F_i^n(t) < 1\}$. $\hat{F}_i$ is the product limit estimator of F_i^n based on the observations $\tilde{X}_{ij}^n$, δ_{ij}^n in the i-th sample.

By Assumption 3.1.1, for each i = 1,...,r, M_i is a zero mean square integrable martingale with

$$(4.1.6) \qquad \langle M_i, M_i \rangle = \int Y_i(1 - \Delta G_i^n)\,dG_i^n$$

$$(4.1.7) \qquad \langle M_i, M_{i'} \rangle = 0 \qquad\qquad i \ne i'.$$

All the processes defined by (4.1.1) to (4.1.5) are adapted; Y_i and J_i are predictable.

By Assumption 3.1.2 (not used in this section), for each t, conditional on $\mathcal{F}_{t-}$, for each i = 1,...,r, $\Delta N_i(t)$ has a binomial distribution with parameters $Y_i(t)$ and $\Delta G_i(t)$. Also, the $\Delta N_i(t)$'s are conditionally independent given $\mathcal{F}_{t-}$.

We shall be particularly interested in the following special case, which includes Examples 3.1.1, 3.1.3 and 3.1.4.

EXAMPLE 4.1.1 "*General random censorship model*".
For each $n = 1,2,\ldots$ X^n_{ij} and U^n_{ij}, $j = 1,\ldots,n_i$, $i = 1,\ldots,r$ are 2n independent positive random variables, X^n_{ij} or U^n_{ij} almost surely finite for each i, j and n. X^n_{ij} has (sub)-distribution function F^n_i and U^n_{ij} has (sub)-distribution function L^n_{ij}. The observable random variables $\tilde{X}^n_{ij}$ and δ^n_{ij} are defined by $\tilde{X}^n_{ij} = X^n_{ij} \wedge U^n_{ij}$, $\delta^n_{ij} = \chi_{\{X^n_{ij} \le U^n_{ij}\}}$.

If (sub)-distribution functions L_1 and L_2 exist such that $L^n_{ij} = L_i$ for all i and n, we speak of the (*usual*) *random censorship model*.

If $L^n_{ij} = \chi_{[u^n_{ij},\infty)}$ for some $u^n_{ij} \in (0,\infty]$, we speak of the *model of fixed censorship*.

We now consider the product limit estimator, setting r = 1 and dropping the index i everywhere. By (3.2.13), if t and n satisfy $F^n(t) < 1$, we have on the event $\{Y(t) > 0\}$

$$(4.1.8) \qquad \frac{\hat{F} - F^n}{1 - F^n} = \int \frac{1 - \hat{F}_-}{1 - F^n} \frac{J}{Y}\, dM \quad \text{on } [0,t].$$

Define

$$(4.1.9) \qquad H = \frac{(1 - \hat{F}_-)J}{(1 - F^n)Y}$$

and

$$(4.1.10) \qquad Z = \int H dM.$$

Again, if t and n satisfy $F^n(t) < 1$, H is a bounded predictable process and M a square integrable martingale on [0,t]. So by (4.1.10) and the theory of stochastic integrals, $Z^2 - \langle Z,Z\rangle$ is a martingale on [0,t], where

$$(4.1.11) \qquad \langle Z,Z\rangle = \int H^2\, d\langle M,M\rangle$$

$$= \int \frac{(1 - \hat{F}_-)^2 J}{(1 - F^n)^2 Y}(1 - \Delta G^n)\, dG^n \qquad ((4.1.6) \text{ and } (4.1.9))$$

is a predictable, nondecreasing, right-continuous process, zero at time zero. By the martingale property and Doob's optional sampling theorem, for all stopping times T < t

$$E(Z(T)^2) = E(\langle Z,Z\rangle(T)).$$

We now see that Theorem 2.4.2 is applicable with Z^2 in the place of X and $\langle Z,Z\rangle$ in the place of Y. The following theorem then becomes straightforward to prove:

THEOREM 4.1.1 *(Consistency of the empirical cumulative hazard function and of the product limit estimator).*
Let $t \in (0,\infty]$ *be such that*

$$Y(t) \to_P \infty \quad \text{as } n \to \infty \tag{4.1.12}$$

and

$$\limsup_{n\to\infty} F^n(t-) < 1. \tag{4.1.13}$$

Then

$$\sup_{s\in[0,t]} |\hat{F}(s)-F^n(s)| \to_P 0 \quad \text{as } n \to \infty \tag{4.1.14}$$

and

$$\sup_{s\in[0,t]} \left| \int_0^s \frac{dN}{Y} - G^n(s) \right| \to_P 0 \quad \text{as } n \to \infty. \tag{4.1.15}$$

If $u \in (0,\infty]$ *is such that* (4.1.12) *and* (4.1.13) *hold for all* $t < u$, *and if furthermore*

$$\lim_{t\uparrow u} \limsup_{n\to\infty} (F^n(u) - F^n(t)) = 0 \tag{4.1.16}$$

then (4.1.14) *holds with the interval* $[0,t]$ *replaced with* $[0,u]$.

PROOF. Letting t be fixed and satisfy (4.1.12) and (4.1.13) we see that

$$P\left(\frac{\hat{F} - F^n}{1 - F^n} = Z \text{ on } [0,t]\right) \to 1 \quad \text{as } n \to \infty,$$

and also

$$\liminf_{n\to\infty} \inf_{s\in[0,t)} (1 - F^n(s)) > 0.$$

So to show first that $\sup_{s\in[0,t)} |\hat{F}(s) - F^n(s)| \to_P 0$ it suffices to show that $\sup_{s\in[0,t)} (Z(s)^2) \to_P 0$. Now by Theorem 2.4.2 applied to the time interval $[0,t)$,

$$P\left(\sup_{s\in[0,t)} Z(s)^2 \geq \varepsilon\right) \leq \frac{\eta}{\varepsilon} + P(\langle Z,Z\rangle(t-) > \eta) \leq$$

$$\leq \frac{\eta}{\varepsilon} + P\left(\frac{G^n(t-)}{(1 - F^n(t-))^2 Y(t)} > \eta\right) \quad \text{(by (4.1.11))}.$$

By (4.1.12) and (4.1.13), the second term on the right hand side converges to zero as $n \to \infty$ for each $\eta > 0$. Since ε and η are arbitrary, we have now shown that

$$\sup_{s\in[0,t)} |\hat{F}(s) - F^n(s)| \to_P 0 \quad \text{as } n \to \infty.$$

By (3.2.7) and (3.2.8), on $\{Y(t) > 0\}$,

$$\Delta\hat{F}(t) - \Delta F^n(t) = (1 - \hat{F}(t-))\frac{\Delta N(t)}{Y(t)} - (1 - F^n(t-))J(t)\Delta G^n(t).$$

So to complete the proof of the first part of the theorem concerning the product limit estimator, we must show that $\Delta N(t)/Y(t) - J(t)\Delta G^n(t) \to_P 0$ as $n \to \infty$. Now since $\int dN/Y - \int JdG^n = \int Y^{-1}dM$ is also a square integrable martingale on $[0,t]$ with $\langle\int Y^{-1}dM, \int Y^{-1}dM\rangle = \int (J/Y)(1 - \Delta G^n)dG^n$, applying Theorem 2.4.2 on the interval $[0,t]$ shows that

$$P\left(\sup_{s\in[0,t]} \left|\int_0^s \frac{dN}{Y} - \int_0^s JdG^n\right| \geq \varepsilon\right) \to 0$$

as $n \to \infty$ for all $\varepsilon > 0$. So this completes the proof that (4.1.14) holds, and also establishes (4.1.15). The rest of the proof is a straightforward monotonicity argument. □

In the situation of Example 4.1.1, we see that

$$EY(t) = (1 - F^n(t-)) \sum_{j=1}^{n} (1 - L_j^n(t-))$$

and

$$\text{var } Y(t) = (1 - F^n(t-)) \sum_{j=1}^{n} \{(1 - L_j^n(t-))(1 - (1-F^n(t-))(1 - L_j^n(t-)))\}$$

$$\leq EY(t).$$

So in this case, and in the presence of Condition (4.1.13), (4.1.12) is equivalent to

$$(4.1.17) \qquad \liminf_{n\to\infty} \sum_{j=1}^{n} (1 - L_j^n(t-)) = \infty.$$

PETERSON (1977), WINTER, FÖLDES & REJTŐ (1978), FÖLDES, REJTŐ & WINTER (1980), and FÖLDES & REJTŐ (1980a) and (1980b) give consistency results under various special cases of Example 4.1.1, under conditions always implying (4.1.13) and (4.1.17). The results of FÖLDES et al. are on strong uniform consistency and include information on rates of convergence. AALEN & JOHANSEN (1978) Theorem 4.5 give the first part of our Theorem 4.1.1 in the case that F^n is independent of n, is continuous, and possesses a hazard rate; otherwise their result is more general as it is concerned with nonparametric estimation of the transition probabilities of a Markov chain.

Actually Theorem 4.1.1 often implicitly gives conditions for uniform consistency of the product limit estimator on the whole real line. For instance, suppose the underlying distribution functions F^n are fixed, $F^n = F$ for all n. As usual, define $\tau = \sup\{t: F(t) < 1\}$. Now (4.1.13) automatically holds for all $t < \tau$, while if $F(\tau-) = F(\tau)$ then (4.1.16) holds. So if (4.1.12) holds with $t = \tau$ in the first case, or for all $t < \tau$ in the second case, uniform consistency is proved on $[0,\tau]$, which is equivalent to uniform consistency on $[0,\infty)$. In this case Theorem 4.1.1 implies consistency of the natural estimator $\int_0^T (1-\hat{F})ds$ of mean lifetime $\int_0^\tau (1-F)ds$. The only difficulty occurs when $\tau = \infty$; but this can be solved, assuming the mean lifetime itself is finite, by using (3.2.22) to bound the tail of the integral by a small finite quantity.

Now we turn to the two-sample tests of the class K^+ of Section 3.3. So in (4.1.1) to (4.1.5), we take $r = 2$. For each $n = n_1+n_2$, $K \in K$ is a bounded predictable process, which is a function of the observations and which is zero where $Y_1 \wedge Y_2$ is zero. If K is nonnegative then we say $K \in K^+$. For convenience we repeat some of the definitions of stochastic processes of Section 3.3 (each defined for each n):

(4.1.18) $$W = \int K \frac{dN_1}{Y_1} - \int K \frac{dN_2}{Y_2}$$

(4.1.19) $$Z = \int \frac{K}{Y_1} dM_1 - \int \frac{K}{Y_2} dM_2 = W - \int K(dG_1^n - dG_2^n)$$

(4.1.20) $$V_1 = \sum_{i=1}^{2} \int \frac{K^2}{Y_i^2}\left(1 - \frac{\Delta N_i - 1}{Y_i - 1}\right) dN_i$$

(4.1.21) $$V_2 = \int \frac{K^2}{Y_1 Y_2}\left(1 - \frac{\Delta N_1 + \Delta N_2 - 1}{Y_1 + Y_2 - 1}\right) d(N_1+N_2).$$

We suppose throughout that $n_1 \wedge n_2 \to \infty$ as $n \to \infty$. A test of the null hypothesis H_0: $F_1^n = F_2^n$ is based on comparing $W(\infty)/\sqrt{V_1(\infty)}$ or $W(\infty)/\sqrt{V_2(\infty)}$ with the standard normal distribution. These test statistics are called test statistics of the class K or K^+, according to whether K is a member of K or K^+. We consider a sequence of one-sided alternative hypotheses and assume that large positive values of the test statistics lead to rejection of H_0.

Throughout the rest of the section we suppose that F_1^n and F_2^n do not depend on n, defining $F_1 = F_1^n$ and $F_2 = F_2^n$ for all n. We define τ_i and G_i, $i = 1,2$, in the usual way. Alternative hypotheses of interest are:

H_1: $\frac{dG_1}{d\mu} \geq \frac{dG_2}{d\mu}$ on $[0,\tau_1 \wedge \tau_2]$ (where μ is a σ-finite measure dominating G_1 and G_2), and $F_1 \neq F_2$.

H_2: $G_1 \geq G_2$ on $[0,\infty)$, and $F_1 \neq F_2$.

H_3: $F_1 \geq F_2$ on $[0,\infty)$, and $F_1 \neq F_2$.

These three types of alternative hypothesis can be called *ordered hazards, ordered cumulative hazards,* and *stochastic ordering* respectively. H_1 implies H_2 and H_3, while if F_1 and F_2 are continuous, H_2 and H_3 are equivalent. The one-sided form of the alternative H_1' given on page 51 is a special case of H_1.

Finally we repeat the definitions of the three test statistics of particular interest, adding standardizing factors depending on n_1 and n_2 only, which loosely speaking keep the variance of $W(\infty)$ bounded away from 0 and ∞ as $n \to \infty$:

$$K_G = \sqrt{\frac{n_1 n_2}{n_1+n_2} \frac{Y_1}{n_1} \frac{Y_2}{n_2}} \tag{4.1.22}$$

$$K_E = \sqrt{\frac{n_1 n_2}{n_1+n_2}} (1-\hat{F}_{1-})(1-\hat{F}_{2-})J_1 J_2 \tag{4.1.23}$$

$$K_C = \sqrt{\frac{n_1 n_2}{n_1+n_2} \frac{Y_1}{n_1} \frac{Y_2}{n_2} \frac{n_1+n_2}{Y_1+Y_2}} . \tag{4.1.24}$$

All are members of $\mathcal{K}^+$.

The following trivial lemma (we omit the proof) splits the proof of consistency into four parts:

LEMMA 4.1.1. *A one-sided test based on* $W(\infty)/\sqrt{V_\ell(\infty)}$ (ℓ = 1 *or* 2) *is consistent against some fixed alternative hypothesis if, under that hypothesis,*

(4.1.25) $Z(\infty)$ *is bounded in probability as* $n \to \infty$

(4.1.26) $V_\ell(\infty)$ *is bounded in probability as* $n \to \infty$

(4.1.27) $V_\ell(\infty)$ *is bounded away from zero in probability as* $n \to \infty$

(4.1.28) $\int_0^\infty K(dG_1 - dG_2) \to_P +\infty$ *as* $n \to \infty$.

Conditions (4.1.25) to (4.1.27), which are true under very weak regularity conditions, are dealt with in the following sequence of lemmas. In the presence of these conditions, (4.1.28) is a necessary and sufficient condition for consistency. Establishing reasonable conditions for (4.1.28) itself will be a trivial enough matter under the alternative hypothesis H_1, but gives a little more trouble under H_2 and H_3.

LEMMA 4.1.2. *Suppose* $\int_0^\infty \frac{K^2}{Y_1}\,dG_1$ *and* $\int_0^\infty \frac{K^2}{Y_2}\,dG_2$ *are bounded in probability as* $n \to \infty$. *Then* (4.1.25) *and* (4.1.26) *with* $\ell = 1$ *hold. If on the other hand* $\int_0^\infty \frac{K^2}{Y_1}\,dG_2$ *and* $\int_0^\infty \frac{K^2}{Y_2}\,dG_1$ *are bounded in probability as* $n \to \infty$, *then* (4.1.26) *holds with* $\ell = 2$.

PROOF. Using (4.1.6), (4.1.7) and the theory of stochastic integrals, we see that the following three processes are all zero-mean martingales on $[0,\infty]$:

$$Z^2 - \sum_i \int \frac{K^2}{Y_i}(1-\Delta G_i)\,dG_i$$

$$\sum_i \int \frac{K^2}{Y_i^2}\,dN_i - \sum_i \int \frac{K^2}{Y_i}\,dG_i$$

and

$$\sum_i \int \frac{K^2}{Y_1 Y_2}\,dN_i - \left(\int \frac{K^2}{Y_2}\,dG_1 + \int \frac{K^2}{Y_1}\,dG_2\right).$$

Note that

$$0 \le V_1 \le \sum_i \int \frac{K^2}{Y_i^2}\,dN_i$$

and that

$$0 \le V_2 \le \sum_i \int \frac{K^2}{Y_1 Y_2}\,dN_i .$$

We now apply Theorem 2.4.2 by using the martingale property of each of the above three processes, to prove (4.1.25) and (4.1.26) with $\ell = 1$ and $\ell = 2$ in turn.

To prove the first set of assertions we make use of the fact that $\sum_i \int_0^\infty \frac{K^2}{Y_i}\,dG_i$ is bounded in probability as $n \to \infty$. By the martingale property, for every stopping time T

$$EZ(T)^2 = E\left(\sum_i \int_0^T \frac{K^2}{Y_i}(1-\Delta G_i)\,dG_i\right) \le E\left(\sum_i \int_0^T \frac{K^2}{Y_i}\,dG_i\right).$$

$\sum_i \int \frac{K^2}{Y_i} dG_i$ is a predictable process. So by Theorem 2.4.2, choosing $T = \infty$ in (2.4.10),

$$P(Z(\infty)^2 \geq C) \leq \frac{\eta}{C} + P\left(\sum_i \int_0^\infty \frac{K^2}{Y_i} dG_i > \eta\right)$$

for any $C > 0$ and $\eta > 0$, because $Z(\infty) = \lim_{t\to\infty} Z(t)$. Since η and C are arbitrary, under the hypothesis of the lemma (4.1.25) follows directly.

The other two cases are proved in exactly the same way. □

LEMMA 4.1.3. *Suppose that there exists a* $t \in \mathbb{R}^+$ *such that for* $i = 1$ *or* 2,

$$0 < F_i(t) < 1$$

$$Y_i(t) \to_P \infty \qquad \textit{as } n \to \infty$$

and

$$\inf_{[0,t]} \frac{K^2}{Y_i} \qquad \textit{is bounded away from zero in probability as } n \to \infty.$$

Then (4.1.27) *holds with* $\ell = 1$.

PROOF. The conditions of the lemma imply that $\sup_{[0,t]} \Delta G_i < 1$ and that $G_i(t) < \infty$. By Theorem 4.1.1, we have

$$\sup_{s\in[0,t]} \left| \int_0^s \frac{dN_i}{Y_i} - G_i(s) \right| \to_P 0 \qquad \text{as } n \to \infty$$

and hence also

$$\sup_{s\in[0,t]} \left| \frac{\Delta N_i(s)}{Y_i(s)} - \Delta G_i(s) \right| \to_P 0 \qquad \text{as } n \to \infty.$$

Since

$$V_1(\infty) \geq \int_0^t \frac{K^2}{Y_i}\left(1 - \frac{\Delta N_i}{Y_i} \cdot \frac{Y_i}{Y_i - 1}\right) \frac{dN_i}{Y_i} \geq$$

$$\geq \inf_{[0,t]} \left(\frac{K^2}{Y_i}\right) \cdot \left(1 - \sup_{[0,t]} \left(\frac{\Delta N_i}{Y_i}\right) \cdot \frac{Y_i(t)}{Y_i(t) - 1}\right) \int_0^t \frac{dN_i}{Y_i}$$

the theorem is proved. □

LEMMA 4.1.4. *Suppose that for* $i = 1$ *or* 2 *there exists* $t \in \mathbb{R}^+$ *such that*

$$0 < F_i(t)$$

$$F_1(t) < 1 \text{ and } F_2(t) < 1$$

$$Y_1(t) \to_P \infty \text{ and } Y_2(t) \to_P \infty \text{ as } n \to \infty$$

and

$$\inf_{s \in [0,t]} \frac{K^2(s)Y_i(s)}{Y_1(s)Y_2(s)} \text{ is bounded away from zero in probability as } n \to \infty.$$

Then (4.1.27) *holds with* $\ell = 2$.

PROOF. The proof is similar to that of Lemma 4.1.3 after writing

$$V_2(\infty) \geq \int_0^t \frac{K^2 Y_i}{Y_1 Y_2}\left(1 - \frac{\Delta N_1}{Y_1}\frac{Y_1}{Y_1+Y_2-1} - \frac{\Delta N_2}{Y_2}\frac{Y_2}{Y_1+Y_2-1}\right)\frac{dN_i}{Y_i}. \qquad \square$$

We now turn to the more important part of Lemma 4.1.1, namely Condition (4.1.28).

LEMMA 4.1.5. *Suppose* $K \in K^+$. *Under* H_1, *if some* $t \in \mathbb{R}^+$ *satisfies both* $G_1(t) > G_2(t)$ *and the conditions of Lemma* 4.1.3, *then* (4.1.28) *holds.*

PROOF. $Y_i(t) \to_P \infty$ as $n \to \infty$ implies that $\inf_{[0,t]} Y_i \to_P \infty$ as $n \to \infty$ and so $\inf_{[0,t]} K \to_P \infty$ as $n \to \infty$. The rest of the proof is now straightforward. $\square$

Before considering the alternative hypotheses H_2 and H_3, we illustrate the previous lemmas by specializing in the following theorem to the test statistics of GEHAN, EFRON and COX. The result is by no means the strongest possible; rather, we have concentrated on making the conditions simple. In particular, the conditions can be weakened if one is only interested in a consistency result with the variance estimator $V_1(\infty)$.

THEOREM 4.1.2 (*Consistency against ordered hazards*).
Consider a fixed alternative in H_1. *Suppose that there exists* $t > 0$ *such that* $G_1(t) > G_2(t)$ *and such that for both* $i = 1$ *and* 2, $0 < F_i(t) < 1$ *and* $Y_i(t)/n_i$ *is bounded away from zero in probability as* $n \to \infty$. *Then* $W_G(\infty)/\sqrt{V_{G\ell}(\infty)} \to_P +\infty$ *as* $n \to \infty$, $\ell = 1$ *and* 2. *Under the additional condition*

$$(4.1.29) \qquad \liminf_{n\to\infty} \frac{n_i}{n_1+n_2} > 0, \quad i = 1 \text{ and } 2,$$

$W_C(\infty)/\sqrt{V_{C\ell}(\infty)} \to_P +\infty$ *as* $n \to \infty$, $\ell = 1$ *and* 2. *Alternatively, under the additional condition that* $Y_i(T)/n_i$ *is bounded away from zero in probability as* $n \to \infty$ *for* $i = 1$ *and* 2, *where* $T = \inf\{s: Y_1(s) \wedge Y_2(s) = 0\}$, $W_E(\infty)/\sqrt{V_{E\ell}(\infty)} \to_P +\infty$ *as* $n \to \infty$, $\ell = 1$ *and* 2.

PROOF. For checking the conditions of Lemma 4.1.2 note that

$$\int \frac{K^2}{Y_i}\, dG_i = n_i \int \frac{K^2}{Y_i^2} \frac{Y_i dG_i}{n_i}$$

and

$$\int \frac{K^2}{Y_1 Y_2} Y_i dG_i = n_i \int \frac{K^2}{Y_1 Y_2} \frac{Y_i dG_i}{n_i},$$

where

$$E\left(\int \frac{Y_i dG_i}{n_i}\right) = E\left(\int \frac{dN_i}{n_i}\right) \le 1.$$

So it suffices to check that

$$\sup_{\mathbb{R}^+} \frac{n_i K^2}{Y_i^2} \quad \text{and} \quad \sup_{\mathbb{R}^+} \frac{n_i K^2}{Y_1 Y_2}$$

are bounded in probability as $n \to \infty$ for each $i = 1,2$ and for each of the three test statistics. For the test statistic of GEHAN, this follows from the relationships

$$\sup_{\mathbb{R}^+} \frac{n_1 K_G^2}{Y_1^2} \le \sup_{\mathbb{R}^+} \frac{n_2}{n_1+n_2}\left(\frac{Y_2}{n_2}\right)^2 \le 1$$

and

$$\sup_{\mathbb{R}^+} n_1 \frac{K_G^2}{Y_1 Y_2} = \sup_{\mathbb{R}^+} \frac{n_1}{n_1+n_2} \frac{Y_1}{n_1} \frac{Y_2}{n_2} \le 1,$$

and those obtained by interchanging the induces 1 and 1. For the test statistic of COX we have similarly

$$\sup_{\mathbb{R}^+} n_1 \frac{K_C^2}{Y_1^2} \le \sup_{\mathbb{R}^+}\left(\frac{Y_2}{Y_1+Y_2}\right)^2 \frac{n_1+n_2}{n_2} \le \frac{n_1+n_2}{n_2}$$

and

$$\sup_{\mathbb{R}^+} n_1 \frac{K_C^2}{Y_1 Y_2} = \sup_{\mathbb{R}^+}\left(\frac{Y_1}{Y_1+Y_2}\right)\left(\frac{Y_2}{Y_1+Y_2}\right) \frac{n_1+n_2}{n_2} \le \frac{n_1+n_2}{n_2}\ .$$

Finally for the test statistic of EFRON we have

$$\sup_{\mathbb{R}^+} n_1 \frac{K_E^2}{Y_1^2} \le \left(\frac{n_1}{Y_1(T)}\right)^2 \frac{n_2}{n_1+n_2}$$

and

$$\sup_{\mathbb{R}^+} n_1 \frac{K_E^2}{Y_1 Y_2} \le \frac{n_1}{Y_1(T)} \frac{n_2}{Y_2(T)} \frac{n_1}{n_1+n_2} .$$

The conditions of Lemmas 4.1.3, 4.1.4 and 4.1.5 are satisfied with the t given by the theorem. Note first that K_G, K_C and K_E are nonincreasing and nonnegative. For such a K,

$$\inf_{[0,t]} \frac{K^2}{Y_i} \ge \frac{K(t)^2}{n_i} .$$

For each test statistic, it is easy to see that if for i = 1 or 2 $\liminf_{n\to\infty} \frac{n_i}{n_1+n_2} > 0$, then for $i' \neq i$, $K(t)^2/n_{i'}$ is bounded away from zero in probability as $n \to \infty$, and so the result is proved in this case. Otherwise, from any subsequence of n's we can extract a further subsequence along which $\liminf \frac{n_i}{n_1+n_2} > 0$ for i = 1 or 2, and so along this sub-sequence $W(\infty)/\sqrt{V_\ell(\infty)} \to_P \infty$. But by a well known result (see e.g. BILLINGSLEY (1968) Theorem 2.3), this implies that $W(\infty)/\sqrt{V_\ell(\infty)} \to_P \infty$ as $n \to \infty$. □

For consistency against more general alternatives we shall have to take more trouble in proving (4.1.28). The next two lemmas will take the place of Lemma 4.1.5 for the alternatives H_2 and H_3. Recall that we have assumed that $n_1 \wedge n_2 \to \infty$ as $n \to \infty$.

LEMMA 4.1.6. *Define* $\tau = \tau_1 \wedge \tau_2$, *and let* k *be a function on* $[0,\infty)$, *zero on* (τ,∞), *such that* $\int_0^\infty |k| dG_i < \infty$, i = 1 *and* 2, *and such that*

$$\int_0^\infty k dG_2 < \int_0^\infty k dG_1 . \tag{4.1.30}$$

Suppose also that $\sqrt{\frac{n_1+n_2}{n_1 n_2}}\, K$ *converges uniformly on* $[0,t]$ *to* k *in probability as* $n \to \infty$ *for each* $t < \tau$, *and that for each* i = 1,2, *either* $G_i(\tau) < \infty$ *and the uniform convergence holds also for* $t = \tau$, *or both*

$$\lim_{t\uparrow\tau} \limsup_{n\to\infty} P\left(\sqrt{\frac{n_1+n_2}{n_1n_2}} \int_{s\in(t,\tau]} |K(s)|\,dG_i(s) < \varepsilon\right) = 0$$

and

$$k(\tau)\Delta G_i(\tau) = 0.$$

Then

$$\int_0^\infty K(dG_1 - dG_2) \to_P +\infty \quad \textit{as } n \to \infty.$$

PROOF. Note that as $n \to \infty$, $\frac{n_1n_2}{n_1+n_2} \to \infty$. Note also that for each n, $K = 0$ on (τ,∞) almost surely. So it suffices to show that as $n \to \infty$

$$\sqrt{\frac{n_1+n_2}{n_1n_2}} \int_0^\tau K\,dG_i \to_P \int_0^\tau k\,dG_i, \quad i = 1 \text{ and } 2.$$

Now by the uniform convergence of $\sqrt{\frac{n_1+n_2}{n_1n_2}}\,K$,

$$\sqrt{\frac{n_1+n_2}{n_1n_2}} \int_0^t K\,dG_i \to_P \int_0^t k\,dG_i, \quad i = 1 \text{ and } 2, \text{ for each } t < \tau,$$

and also for $t = \tau$ if $G_i(\tau) < \infty$ and the uniform convergence holds on $[0,\tau]$. In the other case $\int_0^t k dG_i \to \int_0^\tau k dG_i$ as $t \uparrow \tau$, and we can see directly or apply BILLINGSLEY (1968) Theorem 4.2 to obtain the required result. □

REMARK 4.1.1. Note the precise meaning of uniform convergence on $[0,t]$ of the process $\sqrt{\frac{n_1+n_2}{n_1n_2}}\,K$ to the function k in probability as $n \to \infty$; this is

$$\sup_{s\in[0,t]} \left| \sqrt{\frac{n_1+n_2}{n_1n_2}}\,K(s) - k(s) \right| \to_P 0 \quad \text{as } n \to \infty.$$

LEMMA 4.1.7. *Let* k *be a nonnegative function such that* $\int_0^\infty k dG_i < \infty$, $i = 1$ *and* 2.

(i) *Under* H_2, *if* k *is left continuous and nonincreasing, and such that* $\int_B dk_+ < 0$, *where* B *is the set on which* $G_1 > G_2$, *then* (4.1.30) *holds.*

(ii) *Under* H_3, *if there exists a left continuous nonincreasing function* g *such that*

$$\frac{k}{1 - F_{1-}} \ge g \ge \frac{k}{1 - F_{2-}}$$

and such that $\int_B dg_+ < 0$ *when* B *is the set on which* $F_1 > F_2$, *then* (4.1.30) *holds.*

(In each case, without the condition involving B *it still holds that* $\int_0^\infty k dG_2 \le \int_0^\infty k dG_1$.*)*

PROOF. (i) Writing $\int k dG_i = k_+G_i - \int G_i dk_+$ (note that $G_i(0) = 0$) we see that $k_+(t)G_i(t)$ tends to a finite limit as $t \to \infty$, and that $\int_0^\infty G_i dk_+$ is finite. So

$$\int_0^\infty k(dG_1 - dG_2) = \lim_{t\to\infty} k_+(t)G_1(t) - \lim_{t\to\infty} k_+(t)G_2(t) + \int_0^\infty (G_1 - G_2)dk_+$$
$$\geq \int_B (G_1 - G_2)dk_+$$
$$> 0.$$

(ii) $$\int_0^\infty k dG_i = \int_0^\infty \frac{k}{1 - F_{i-}} dF_i.$$

So $\int_0^\infty k(dG_1 - dG_2) \geq \int_0^\infty g(dF_1 - dF_2) > 0$ by the same arguments used to prove (i). □

Combining the conditions of Theorem 4.1.2 with those of Lemmas 4.1.6 and 4.1.7 gives consistency results for the test statistics of COX, GEHAN and EFRON against alternatives H_2 and H_3. In the first two cases, uniform convergence of $\sqrt{\frac{n_1+n_2}{n_1 n_2}}\, K$ to a function k as $n \to \infty$ is difficult to imagine without uniform convergence of Y_1/n_1 and Y_2/n_2 to functions y_1 and y_2 say. Note that such functions y_i are necessarily nonincreasing, nonnegative, left continuous and even such that $y_i/(1 - F_{i-})$ is nonincreasing. For $Y_i/(1 - \hat{F}_{i-})$ is nonincreasing (see the remarks following Definition (3.2.6)), so for $s < t$

$$\frac{Y_i(t)}{Y_i(s)} \leq \frac{1 - \hat{F}_i(t-)}{1 - \hat{F}_i(s-)} \to_P \frac{1 - F_i(t-)}{1 - F_i(s-)}, \quad \text{if } y_i(t) > 0,$$

by Theorem 4.1.1. This makes the following theorem easy to prove:

THEOREM 4.1.3 *(Consistency of the test statistics of GEHAN and COX against ordered cumulative hazards or stochastic ordering).*
Consider a fixed alternative in H_2 *or* H_3. *Suppose functions* y_1 *and* y_2 *exist such that* Y_i/n_i *converges uniformly on* $[0,\infty)$ *to* y_i *in probability as* $n \to \infty$, $i = 1,2$.
Suppose a $t > 0$ *exists such that for* $i = 1$ *and* 2, $0 < F_i(t) < 1$ *and* $y_i(t) > 0$. *Then* y_1 *and* y_2 *satisfy*

$$(4.1.31)\qquad \int_0^\infty y_1y_2(dG_1-dG_2) \ge 0$$

and

$$(4.1.32)\qquad \int_0^\infty \frac{y_1y_2}{\rho_1y_1+\rho_2y_2}(dG_1-dG_2) \ge 0 \qquad (0 < \rho_1 < 1,\ \rho_1+\rho_2 = 1)$$

hold. If (4.1.31) *is strict, then* $W_G(\infty)/\sqrt{W_{G\ell}(\infty)} \to_P +\infty$ *as* $n \to \infty$, $\ell = 1$ *and* 2, *while if* $\liminf_{n\to\infty} \frac{n_i}{n_1+n_2} > 0$, $i = 1$ *and* 2, *and* (4.1.32) *is strict for all limit points* (ρ_1,ρ_2) *of* $(\frac{n_1}{n_1+n_2}, \frac{n_2}{n_1+n_2})$, *then* $W_C(\infty)/\sqrt{V_{C\ell}(\infty)} \to_P +\infty$ *as* $n \to \infty$, $\ell = 1$ *and* 2.

PROOF. Under the conditions of this theorem, all the conditions of Theorem 4.1.2 hold, with the single exception of the condition $G_1(t) > G_2(t)$ for the right t. However this condition was only needed to make Lemma 4.1.5 applicable, with which we proved (4.1.28). So it only remains to prove (4.1.28), for which we shall use Lemmas 4.1.6 and 4.1.7. Defining $k_G = y_1y_2$ and $k_C = \frac{y_1y_2}{\rho_1y_1+\rho_2y_2}$, we see that k_G and k_C are nonnegative, left continuous and nonincreasing (by the remarks preceding the Theorem). Also we see that $\int_0^\infty k_G dG_i \le \int_0^\infty y_i dG_i \le \int_0^\infty (1-F_{i-})dG_i \le 1$ and that $\int_0^\infty k_C dG_i \le \rho_{i'}^{-1}\int_0^\infty y_i dG_i \le$ $\le \rho_{i'}^{-1}$ $(i \ne i')$. So (4.1.31) and (4.1.32) hold under H_2 by the last line of Lemma 4.1.7.

For H_3, note that $k_G(1-F_{i-})^{-1}$ is nonincreasing and left continuous, and that $k_G(1-F_{1-})^{-1} \ge k_G(1-F_{2-})^{-1}$, so we can choose g to be either of these functions in applying the second part of Lemma 4.1.7 to k_G. Similarly we have under H_3

$$\begin{aligned}
\frac{k_C}{(1-F_{1-})} &= \Big((1-F_{1-})(\rho_1y_2^{-1} + \rho_2y_1^{-1})\Big)^{-1} \\
&= \left(\rho_1\frac{1-F_{1-}}{y_2} + \rho_2\frac{1-F_{1-}}{y_1}\right)^{-1} \\
&\ge \left(\rho_1\frac{1-F_{2-}}{y_2} + \rho_2\frac{1-F_{1-}}{y_1}\right)^{-1} \\
&\ge \left(\rho_1\frac{1-F_{2-}}{y_2} + \rho_2\frac{1-F_{2-}}{y_1}\right)^{-1} \\
&= \Big((1-F_{2-})(\rho_1y_2^{-1} + \rho_2y_1^{-1})\Big)^{-1} = \frac{k_C}{(1-F_{2-})},
\end{aligned}$$

where the central expression in the chain is a left continuous nonincreasing function. So (4.1.31) and (4.1.32) also hold under H_3.

It remains to verify the conditions on the convergence of $\sqrt{\frac{n_1+n_2}{n_1n_2}}\,K$ in Lemma 4.1.6. For the test statistic of GEHAN we have that $\sqrt{\frac{n_1+n_2}{n_1n_2}}\,K_G$ converges uniformly on $[0,\infty)$ to k_G in probability as $n \to \infty$. If for $i = 1$ or 2, $G_i(\tau) = \infty$, then $k_G(\tau)\Delta G_i(\tau) = 0$ and

$$E\left(\sqrt{\frac{n_1+n_2}{n_1n_2}}\int_{s\in(t,\tau]} K_G(s)dG_i(s)\right) \le E\left(\int_{s\in(t,\tau]} \frac{Y_i(s)dG_i(s)}{n_i}\right)$$

$$= E\left(\frac{N_i(\tau) - N_i(t)}{n_i}\right)$$

$$\le F_i(\tau) - F_i(t) \to 0 \quad \text{at } t \uparrow \tau$$

uniformly in n. So the conditions of Lemma 4.1.6 are satisfied for $K = K_G$.

For the test statistic of COX, suppose first that $\frac{n_i}{n_1+n_2} \to \rho_i \in (0,1)$ as $n \to \infty$. Then we certainly have that $\sqrt{\frac{n_1+n_2}{n_1n_2}}\, K_C$ converges uniformly on $[0,u]$ to k_C in probability as $n \to \infty$ for each u such that $y_i(u) > 0$, $i = 1,2$. Since $\sqrt{\frac{n_1+n_2}{n_1n_2}}\, K_C \le \frac{Y_i}{n_i}\,\frac{(n_1+n_2)n_i}{n_1n_2}$ and $k_C \le y_i\,\frac{\rho_i}{\rho_1\rho_2}$, it is easy to see that the convergence can be extended to $[0,\infty)$.

If for $i = 1$ or 2, $G_i(\tau) = \infty$, then $K_C(\tau)\Delta G_i(\tau) = 0$ and

$$E\left(\sqrt{\frac{n_1+n_2}{n_1n_2}}\int_{s\in(t,\tau]} K_C(s)dG_i(s)\right)$$

$$\le \frac{(n_1+n_2)n_i}{n_1n_2}\, E\left(\int_{s\in(t,\tau]} \frac{Y_i(s)dG_i(s)}{n_i}\right) \to 0 \quad \text{at } t \uparrow \tau$$

uniformly in n; which completes the proof of the theorem when $\frac{n_i}{n_1+n_2}$ converges as $n \to \infty$. Otherwise, for any subsequence we can extract a further subsequence along which $\lim \frac{n_i}{n_1+n_2} = \rho_i$ for some $\rho_i \in (0,1)$. For this sub-subsequence we have $W_C(\infty)/\sqrt{V_{C\ell}(\infty)} \to_P +\infty$; and so the result holds in general. □

We now prove a similar result for the test statistic of EFRON:

<u>THEOREM 4.1.4</u> *(Consistency of the test statistic of EFRON against ordered cumulative hazards or stochastic ordering).*
Consider a fixed alternative in H_2 *or* H_3*. Define* $T = \inf\{s\colon Y_1(s) \wedge Y_2(s) = 0\}$ *and suppose that* $Y_i(T)/n_i$ *is bounded away from zero in probability as* $n \to \infty$ *for* $i = 1$ *and* 2*. Suppose there exists* $t > 0$ *such that* $P(T \ge t) \to 1$ *as* $n \to \infty$ *and such that* $0 < F_i(t) < 1$, $i = 1$ *and* 2*, and suppose there exists a set* B *such that* $P(T \in B) \to 1$ *as* $n \to \infty$ *and*

$$\inf_{s\in B}\left(\int_0^s (1-F_{2-})dF_1 - \int_0^s (1-F_{1-})dF_2\right) > 0$$

(the function $\int (1-F_{2-})dF_1 - \int (1-F_{1-})dF_2$ *is automatically nonnegative). Then* $W_E(\infty)/\sqrt{V_{E\ell}(\infty)} \to_P +\infty$ *as* $n \to \infty$, $\ell = 1$ *and* 2.

PROOF. As in the proof of Theorem 4.1.3, we only have to supply a proof of (4.1.28). Now

$$\frac{Y_i(T)}{n_i} \le \frac{\#\{j: X_{ij} \ge T\}}{n_i}\,.$$

So by the Glivenko-Cantelli theorem, for each $\varepsilon > 0$

$$P\left(\frac{Y_i(T)}{n_i} \le 1-F_i(T-)+\varepsilon\right) \to 1 \quad \text{as } n \to \infty.$$

By the hypothesis of the Theorem, $F_i(T-)$ is bounded away from 1 in probability as $n \to \infty$, $i = 1$ and 2. Now because T is a stopping time it is possible to repeat the proof of the first part of Theorem 4.1.1 with t replaced everywhere with T (in particular, in (4.1.12), (4.1.13) and (4.1.14)). So

$$\sqrt{\frac{n_1+n_2}{n_1n_2}}\,K_E - (1-F_{1-})(1-F_{2-})J_1J_2$$

converges uniformly on $[0,\infty)$ in probability to zero as $n \to \infty$. Because $F_i(T-)$ is bounded away from 1 in probability as $n \to \infty$, $G_i(T)$ is bounded away from ∞, and so

$$\sqrt{\frac{n_1+n_2}{n_1n_2}}\int_0^\infty K_E(dG_1-dG_2) - \int_0^T (1-F_{1-})(1-F_{2-})(dG_1-dG_2)$$

converges in probability to zero as $n \to \infty$. But (4.1.28) follows now immediately because

$$\int (1-F_{1-})(1-F_{2-})(dG_1-dG_2) = \int (1-F_{2-})dF_1 - \int (1-F_{1-})dF_2.$$

It can be seen that this function is nonnegative under H_2 or H_3 by applying Lemma 4.1.7. □

We conclude this section with some remarks on Theorems 4.1.2 to 4.1.4. Note first of all that for the test statistic of COX we made the assumption that $\liminf_{n\to\infty} \frac{n_i}{n_1+n_2} > 0$ for $i = 1$ and 2. This assumption can certainly be dropped in many situations but only at the cost of a far more complicated proof; we shall go into this matter more deeply when proving asymptotic normality in Section 4.3, when the same problem arises.

For the test statistic of EFRON we imposed the rather strong condition that $Y_i(T)/n_i$ is bounded away from zero in probability as $n \to \infty$, where $T = \inf\{s: Y_1(s) \wedge Y_2(s) = 0\}$. However, as we shall see in the next section and as EFRON (1967) remarked, his test statistic will often fail to be asymptotically normally distributed, unless one is prepared to use not $W_E(\infty)$ but $W_E(t)$ as a test statistic, where t is such that for i = 1 and 2 $Y_i(t)/n_i$ converges in probability to a positive quantity as $n \to \infty$. So our condition is not restrictive at all if one follows this advice; t can even be replaced with a stopping time. Note also that by Theorem 4.1.4 his test statistic seems particularly suited to testing H_0 against the alternative hypothesis

$$H_4: \quad P(X_1 \le X_2 \wedge t) \ge P(X_2 \le X_1 \wedge t) \quad \text{for all } t,$$

where X_1 and X_2 are independently distributed with distribution functions $F_1 \neq F_2$. If F_1 and F_2 are continuous, H_4 is equivalent to $P(X_1 \wedge t \le X_2 \wedge t) \ge$ $\ge P(X_2 \wedge t \le X_1 \wedge t)$ for all t. As we saw (Lemma 4.1.7), H_4 is implied by both H_2 and H_3.

In Example 4.1.1, a sufficient condition for convergence of Y_i/n_i is

$$(4.1.33) \qquad \frac{1}{n_i} \sum_{j=1}^{n_i} L_{ij}^n(t) \to L_i(t) \quad \text{uniformly in } t \in [0,\infty)$$

as $n \to \infty$ for some (sub)-distribution functions L_i, i = 1 and 2. This can be shown by applying the Glivenko-Cantelli theorem for independent but not necessarily identically distributed random variables of VAN ZUIJLEN (1978) (see his Theorem 2.1, Remark 2.1 and Corollary 3.1). In this case, $y_i = (1-F_{i-})(1-L_{i-})$.

Note that in Example 4.1.1,

$$\operatorname{var}(Y_i(t)) \le EY_i(t) = (1-F_i(t-)) \sum_{j=1}^{n_i} (1-L_{ij}^n(t-))$$

and

$$EY_i(t+) = (1-F_i(t)) \sum_{j=1}^{n_i} (1-L_{ij}^n(t)).$$

So in this case the condition in Theorem 4.1.4 involving $Y_i(T)/n_i$ could be replaced with the following one:

"There exists $t > 0$ such that $F_i(t-) < 1$, $i = 1$ and 2, such that for each n and for $i = 1$ or 2, $(1 - F_i(t)) \sum_{j=1}^{n_i} (1 - L_{ij}^n(t)) = 0$, and such that $\liminf_{n\to\infty} \frac{1}{n_i} \sum_{j=1}^{n_i} (1 - L_{ij}^n(t-)) > 0$, $i = 1$ and 2."

Under this condition $P(T=t) \to 1$ as $n \to \infty$.

Results on Example 3.1.2 and similar cases can be easily obtained by adapting the approach used above as follows. Let K, Y_i, N_i, etc. be the usual processes which correspond to the experiment described in Example 3.1.2 when the experiment is *not* terminated at some predetermined failure, but allowed to continue indefinitely. Then the test statistic corresponding to the *stopped* experiment is $W(T)/\sqrt{V_\ell(T)}$, $\ell = 1$ or 2, where T is some stopping time. Equivalently, stopping the experiment corresponds to replacing K with $K \cdot \chi_{[0,T]}$, which is also a predictable process having all the usual properties if T is a stopping time depending on observable quantities.

Now the conditions of Lemma 4.1.6 in fact ensure that $\sqrt{\frac{n_1+n_2}{n_1 n_2}} \int KdG_i$ converges uniformly on $[0,\infty)$ to the function $\int kdG_i$ in probability as $n \to \infty$, for each $i = 1,2$, so we can conclude that

$$(4.1.34) \qquad \int_0^T K(dG_1 - dG_2) \to_P +\infty$$

as $n \to \infty$, if there exists a set B such that $P(T \in B) \to 1$ as $n \to \infty$ and $\inf_B (\int k(dG_1 - dG_2)) > 0$. But (4.1.34) is exactly (4.1.28) if K is replaced with $K\chi_{[0,T]}$ in the latter. Again (4.1.25) to (4.1.27) with $Z(\infty)$ and $V_\ell(\infty)$ replaced with $Z(T)$ and $V_\ell(T)$ will hold under very weak regularity conditions.

4.2. Weak convergence: general theorem and the product limit estimator

This section contains a general weak convergence theorem. As an application we prove weak convergence of the product limit estimator and use the result to construct confidence bands for an unknown distribution function F. In Section 4.3 we shall apply the general theorem in the two-sample case, to derive conditions under the null hypothesis for a test statistic of the class $\mathcal{K}$ to be asymptotically normally distributed. Our general theorem, Theorem 4.2.1, is an adaptation of Theorem 2.4.1 to the situation described at the beginning of Section 4.1: a sequence (as $n = 1,2,\ldots$) of r-sample set-ups with a total of $n = \sum_{i=1}^r n_i$ observations $(\tilde{X}_{ij}^n, \delta_{ij}^n)$, $j = 1,\ldots,n_i$,

$i = 1,\ldots,r$. The notation here will be exactly as in Section 4.1, so that in particular dependence on n will be suppressed, except as far as the underlying distribution functions F_i^n and the associated functions $G_i^n = \int (1 - F_{i-}^n)^{-1} dF_i^n$ are concerned (we allow F_i^n to depend on n so as to be able to deal with a contiguous sequence of alternative hypotheses in our discussion of efficiencies in Chapter 5).

Theorem 4.2.1 gives conditions for joint weak convergence of processes $Z_i = \int H_i dM_i$ where for each n, M_i is the square integrable martingale defined by (4.1.3), and H_i is a bounded predictable process. So for the product limit estimator (Theorem 4.2.2), H_i will be defined by (4.1.9) (where the index i has been dropped because $r = 1$), and for two-sample tests of the class K (Corollaries 4.3.1 and 4.3.2) H_i is defined to be K/Y_i (see (4.1.19) for the general case, and (4.1.22) to (4.1.24) for the special case of the test statistics of GEHAN, EFRON and COX). Corollaries 4.3.1 and 4.3.2 are in fact little more than this substitution of K/Y_i for H_i in the conditions of Theorem 4.2.1. However in Propositions 4.3.1 to 4.3.3 we verify these conditions in a very general situation for the test statistics of GEHAN, COX and EFRON. We close Section 4.3 with a discussion of these results.

We take as given the situation specified at the beginning of Section 4.1, so that in particular Assumptions 3.1.1 and 3.1.2 hold. Let us start by stating a list of conditions. Here, I is the interval $[0,u)$ or $[0,u]$ for some fixed $u \in (0,\infty]$, F_i is some fixed (sub)-distribution function and $G_i = \int (1 - F_{i-})^{-1} dF_i$, $i = 1,\ldots,r$. For each i, h_i is a nonnegative function finite on I and zero outside I.

I. For each $i = 1,\ldots,r$

a) F_i^n converges uniformly on I to F_i as $n \to \infty$; G_i is finite on I.

b) $H_i^2 Y_i$ converges uniformly on each closed subinterval of I in probability to h_i as $n \to \infty$; h_i is left continuous with right hand limits and h_{i+} of bounded variation on each closed subinterval of I if F_i^n varies with n; if F_i^n is fixed, h_i need only be bounded on each closed subinterval of I.

c) $Y_i(t) \to_P \infty$ as $n \to \infty$ for each $t \in I$.

II. If $u \notin I$, then for each $i = 1,\ldots,r$

a) $\int_I h_i (1 - \Delta G_i) dG_i < \infty$.

b) $\lim_{t \uparrow u} \limsup_{n \to \infty} P(\int_{(t,u]} H_i^2 Y_i dG_i^n > \varepsilon) = 0$ for all $\varepsilon > 0$.

III. If $u < \infty$, then for each $i = 1,\ldots,r$

$\int_{(u,\infty)} H_i^2 Y_i dG_i^n \to_P 0$ as $n \to \infty$.

<u>THEOREM 4.2.1</u>. *Suppose that for each* n, $H_1,\ldots,H_r$ *are bounded predictable processes, and define square integrable martingales* $Z_i = \int H_i dM_i$. *Suppose that Condition* I *holds for some* $I = [0,u)$ *or* $[0,u]$ *and some functions* h_i, *and let* $Z_1^\infty,\ldots,Z_r^\infty$ *be independent zero mean Gaussian processes with independent increments and variance functions* $\int h_i(1-\Delta G_i)dG_i$, *defined on* I. *If Condition* II *holds, such processes are also defined on* $[0,\infty]$. *Then*

$$\{Z_i: i = 1,\ldots,r\} \to_{\mathcal{D}} \{Z_i^\infty: i = 1,\ldots,r\} \quad \text{as } n \to \infty$$

in $(D(I))^r$, *and a Skorohod-type construction (see Theorem* 2.4.3) *is possible with* $\sup_{s\in[0,t]} |Z_i(s) - Z_i^\infty(s)| \to 0$ *as* $n \to \infty$ *almost surely for each* $t \in I$ *and each* $i = 1,\ldots,r$. *Adding Condition* II, *this statement also holds with* I *replaced everywhere by* $[0,u]$, *and also adding* III, *with* I *replaced with* $[0,\infty]$.

<u>PROOF</u>. We may suppose throughout that Condition I holds. By Ia and Ib, and using the fact that G_i is finite on I, it is easy to show that $\langle Z_i,Z_i\rangle = \int H_i^2 Y_i(1-\Delta G_i^n)dG_i^n$ converges uniformly on $[0,t]$ to $\int h_i(1-\Delta G_i)dG_i$ in probability as $n \to \infty$, for each $t \in I$. If Condition II holds too, then arguing directly or by BILLINGSLEY (1968) Theorem 4.2, we have uniform convergence on $[0,u]$; adding Condition III extends this to uniform convergence on $[0,\infty]$. Moreover, for $i \neq i'$, $\langle Z_i,Z_{i'}\rangle = 0$ for all n.

Next, for each $\varepsilon > 0$, for each n and each $i = 1,\ldots,r$, define processes J_ε and $R_{i\varepsilon}$ on $[0,\infty)$ by

$$J_\varepsilon(t) = \chi_{\{|H_i(t)|\le\varepsilon, i=1,\ldots,r\}}$$

and

$$R_{i\varepsilon} = \int H_i^2 Y_i(1-J_\varepsilon)(1-\Delta G_i^n)dG_i^n = \int (1-J_\varepsilon)d\langle Z_i,Z_i\rangle.$$

Note that J_ε is predictable and that

$$\sup_{s\in[0,t]} H_i^2(s) \le \frac{\sup_{s\in[0,t]} |H_i^2(s)Y_i(s)|}{\inf_{s\in[0,t]} Y_i(s)}$$

$$\le \frac{\sup_{s\in[0,t]} h_i(s) + \sup_{s\in[0,t]} |H_i^2(s)Y_i(s) - h_i(s)|}{Y_i(t)}.$$

So by Ib and Ic, $\sup_{s\in[0,t]} H_i^2(s) \to_P 0$ as $n \to \infty$ for each $t \in I$, consequently for each i, $\varepsilon > 0$ and $t \in I$,

$$P(R_{i\varepsilon}(t) = 0) \to 1 \quad \text{as } n \to \infty.$$

This certainly implies that $\sup_{s\in[0,t]} R_{i\varepsilon}(s) \to_P 0$ as $n \to \infty$, for each i, $\varepsilon > 0$, and $t \in \mathcal{I}$. Adding Condition IIb extends this to $t = u$, and adding Condition III as well extends it to all $t \in [0,\infty]$.

For each n = 1,2,... and each i = 1,...,r, define

$$\underline{Z}_i^\varepsilon = \int J_\varepsilon dZ_i = \int J_\varepsilon H_i dM_i$$

and

$$\bar{Z}_i^\varepsilon = Z_i - \underline{Z}_i^\varepsilon = \int (1-J_\varepsilon)H_i dM_i.$$

Note that for any i, i' and ε

$$\sup_{[0,\infty]} |\Delta \underline{Z}_i^\varepsilon| \le \varepsilon \sup_{[0,\infty]} |\Delta M_i|,$$

$\bar{Z}_i^\varepsilon$ and $\underline{Z}_{i'}^\varepsilon$ never jump simultaneously, and

$$\langle \bar{Z}_i^\varepsilon, \bar{Z}_i^\varepsilon \rangle = R_{i\varepsilon}.$$

If F_i^n is continuous for all i and n (and so F_i is continuous for all i too) then almost surely,

$$\sup_{s\in[0,\infty]} |\Delta M_i(s)| \le 1$$

for each i and n, and $\int h_i(1 - \Delta G_i)dG_i$ is a continuous function. Theorems 2.4.1 and 2.4.3 now immediately give all the required conclusions.

Suppose on the other hand that some or all of the F_i^n's and F_i's have discontinuities. We can at least enumerate all these discontinuities in a single sequence $t_1, t_2, \ldots$, say. The idea of the proof will be to spread the jump that N_i makes at t_m over a time interval which will be inserted at this point. After this is done, and all the other processes are suitably defined over the inserted intervals, Theorem 2.4.1 will apply giving a continuous process in the limit. Then by deleting all the new time intervals, we shall obtain the required result.

Choose $\delta_m > 0$, m = 1,2,..., such that $\sum_{m=1}^\infty \delta_m < \infty$. Define the time transformation $\phi^*: [0,\infty] \to [0,\infty]$ by

$$\phi^*(t) = t + \sum_{m: t_m \le t} \delta_m.$$

Define $\delta(t) = \Delta\phi^*(t)$. So $\delta(t) = \delta_m$ if $t = t_m$ for some m, otherwise $\delta(t) = 0$. Let $I^* = [0,\phi^*_-(u))$ if $u \notin I$ and $I^* = [0,\phi^*(u)]$ if $u \in I$. Note that for each t^* there exists a unique t such that $\phi^*_-(t) \le t^* \le \phi^*(t)$, and $t \in I$ if $t^* \in I^*$.

We define processes N_i^*, Y_i^*, M_i^*, Z_i^*, H_i^*, J_ε^*, $Z_{-i}^{\varepsilon *}$ and $\bar{Z}_i^{\varepsilon *}$ on the extended time axis as follows. Firstly, if $t^* = \phi^*(t)$ for some t, we define $N_i^*(t^*) = N_i(t)$, etc. Next, extending (Ω,F,P) if necessary, we define N_i^* on the interval $[\phi^*_-(t_m),\phi^*(t_m))$ by letting N_i^* make, conditional on $Y_i(t_m)$ and $\Delta N_i(t_m)$, $\Delta N_i(t_m)$ jumps of size +1 at a random selection of $\Delta N_i(t_m)$ points out of the $Y_i(t_m)$ points

$$\phi^*_-(t_m) + \frac{\ell}{Y_i(t_m)+1}\,\delta_m, \qquad \ell = 1,\ldots,Y_i(t_m).$$

This is done independently over all i and m. We let Y_i^* and H_i^* be equal to $Y_i^*(\phi^*(t_m)) = Y_i(t_m)$ and $H_i^*(\phi^*(t_m)) = H_i(t_m)$ respectively on the interval $[\phi^*_-(t_m),\phi^*(t_m))$; and for $t^* \in [\phi^*_-(t_m),\phi^*(t_m))$ we define

$$M_i^*(t^*) = M_i(t_m-) + (N_i^*(t^*) - N_i(t_m-))$$

$$- \left[(Y_i(t_m)+1)\,\frac{t^*-\phi^*_-(t_m)}{\delta_m}\right] \Delta G_i^n(t_m).$$

(We write $[x]$ for the entier of x.) So M_i^* is piecewise constant on this interval with jumps of size $\Delta N_i^*(t^*) - \Delta G_i^n(t_m)$ at the $Y_i(t_m)$ points defined above. Now conditional on F_{t_m-}, $Y_i(t_m)$ is fixed and $\Delta N_1(t_m),\ldots,\Delta N_r(t_m)$ are independent, $\Delta N_i(t_m)$ being binomially distributed with parameters $Y_i(t_m)$ and $\Delta G_i^n(t_m)$. So conditional on F_{t_m-}, N_i^* makes independently over $i=1,\ldots,r$ and $\ell = 1,\ldots,Y_i(t_m)$ a jump of size +1 at the point $\phi^*_-(t_m) + \frac{\ell}{Y_i(t_m)+1}\delta_m$ with probability $\Delta G_i^n(t_m)$.

Next define σ-algebras $F^*_{t^*}$ by

$$F^*_{t^*} = \begin{cases} F_t \vee \sigma\{N_i^*(s^*)\colon s^* \le t^*,\ i=1,\ldots,r\} & \text{if } t^* = \phi^*(t), \\ F_{t-} \vee \sigma\{N_i^*(s^*)\colon s^* \le t^*,\ i=1,\ldots,r\} & \text{if } \phi^*_-(t) \le t^* < \phi^*(t). \end{cases}$$

We now see that M_i^*, $i=1,\ldots,r$ is a square integrable martingale with respect to $\{F^*_{t^*}\colon t^* \in [0,\infty]\}$, with

$$\langle M_i^*,M_{i'}^*\rangle = 0, \qquad i \ne i'$$

and

$$\langle M_i^*, M_i^* \rangle(t^*) = \begin{cases} \langle M_i, M_i \rangle(t) & \text{if } t^* = \phi^*(t), \\ \langle M_i, M_i \rangle(t-) + \left[(Y_i(t)+1) \dfrac{t^* - \phi_-^*(t)}{\delta(t)} \right] (1 - \Delta G_i^n(t)) \Delta G_i^n(t) & \text{if } \phi_-^*(t) \le t^* < \phi^*(t). \end{cases}$$

We can define $Z_i^*(t^*) = \int_0^{t^*} H_i^* dM_i^*$ for all t^*. Note that H_i^* and Y_i^* are predictable with respect to $\{F_{t^*}^*: t^* \in [0,\infty]\}$, so that Z_i^* is a square integrable martingale for each i. We define as previously

$$J_\varepsilon^*(t^*) = \chi_{\{H_i^*(t^*) \le \varepsilon, i=1,\dots,r\}}$$

$$\underline{Z}_i^{\varepsilon*} = \int J_\varepsilon^* dZ_i^* \quad \text{and} \quad \bar{Z}_i^{\varepsilon*} = Z_i^* - \underline{Z}_i^{\varepsilon*}.$$

Note that for any i and $\varepsilon > 0$, almost surely

$$\sup_{t^* \in [0,\infty]} |\Delta \underline{Z}_i^{\varepsilon*}(t^*)| \le \varepsilon \sup_{t^* \in [0,\infty]} |\Delta M_i^*(t^*)| \le \varepsilon.$$

Also with probability 1, $\underline{Z}_i^{\varepsilon*}$ and $\bar{Z}_{i'}^{\varepsilon*}$ never jump simultaneously for all i, i' and $\varepsilon > 0$,

$$\langle Z_i^*, Z_{i'}^* \rangle = 0 \quad \text{for all } i \ne i'$$

and if $\phi_-^*(t) \le t^* \le \phi^*(t)$, then

$$\langle \bar{Z}_i^{\varepsilon*}, \bar{Z}_i^{\varepsilon*} \rangle(t^*) \le \langle \bar{Z}_i^{\varepsilon}, \bar{Z}_i^{\varepsilon} \rangle(t) \to_P 0$$

as $n \to \infty$, as long as $t^* \in I^*$. If Condition II holds, this is also true for $t^* \in [0,\phi^*(u)]$, while under the further addition of Condition III, even for $t^* \in [0,\infty]$.

So to apply Theorem 2.4.1 to $\{Z_i^*: i = 1,\dots,r\}$, it remains to show that $\langle Z_i^*, Z_i^* \rangle(t^*)$ converges in probability to some continuous function as $n \to \infty$ for each $t^* \in I^*$, $[0,\phi^*(u)]$ or $[0,\infty]$ according to whether Conditions I, I and II, or I, II and III hold.

Now if $t^* = \phi^*(t)$ then $\langle Z_i^*, Z_i^* \rangle(t^*) = \langle Z_i, Z_i \rangle(t) \to_P \int_0^t h_i(1 - \Delta G_i) dG_i$ under the appropriate set of conditions. If however $\phi_-^*(t) \le t^* < \phi^*(t)$, then

$$\langle Z_i^*,Z_i^*\rangle(t^*) =$$

$$\langle Z_i,Z_i\rangle(t-) + H_i^2(t)Y_i(t)\frac{\left[(Y_i(t)+1)\frac{t^*-\phi_-^*(t)}{\delta(t)}\right]}{Y_i(t)}(1-\Delta G_i^n(t))\Delta G_i^n(t).$$

According to whether $t^* \in I^*$, $[0,\phi^*(u)]$ or $[0,\infty]$ we have $t \in I$, $[0,u]$ or $[0,\infty]$ respectively. In each case, under the relevant set of conditions,

$$\langle Z_i,Z_i\rangle(t-) \to_P \int_0^{t-} h_i(1-\Delta G_i)dG_i \quad \text{as } n \to \infty.$$

If $t \in I$, then by Ic, $Y_i(t) \to_P \infty$ and so

$$\left[(Y_i(t)+1)\frac{t^*-\phi_-^*(t)}{\delta(t)}\right] \Big/ Y_i(t) \to_P \frac{t^*-\phi_-^*(t)}{\delta(t)} \quad \text{as } n \to \infty.$$

By Ib, $H_i^2(t)Y_i(t) \to_P h_i(t)$ and by Ia, $(1-\Delta G_i^n(t))\Delta G_i^n(t) \to (1-\Delta G_i(t))\Delta G_i(t)$. So for $t^* \in I^*$, $\phi_-^*(t) \le t^* \le \phi^*(t)$,

$$\text{(4.2.1)} \qquad \langle Z_i^*,Z_i^*\rangle(t^*) \to_P \int_0^{t-} h_i(1-\Delta G_i)dG_i + \frac{t^*-\phi_-^*(t)}{\delta(t)}h_i(t)(1-\Delta G_i(t))\Delta G_i(t)$$

as $n \to \infty$.

If $u \notin I$ and II holds, then using the convergence of $\langle Z_i^*,Z_i^*\rangle$ on I^* that has just been proved and using BILLINGSLEY (1968) Theorem 4.2 in the same way as before, we see that

$$\langle Z_i^*,Z_i^*\rangle(\phi_-^*(u)) \to_P \int_0^{u-} h_i(1-\Delta G_i)dG_i.$$

Also by IIb, for each $\varepsilon > 0$,

$$\limsup_{n\to\infty} P(H_i^2(u)Y_i(u)(1-\Delta G_i^n(u))\Delta G_i^n(u) > \varepsilon) = 0,$$

which implies that

$$\langle Z_i^*,Z_i^*\rangle(\phi^*(u)) - \langle Z_i^*,Z_i^*\rangle(\phi_-^*(u)) \to_P 0.$$

Thus under the addition of II, (4.2.1) holds for all $t^* \in [0,\phi^*(u)]$.

Finally, if III holds as well, then

$$\langle Z_i^*,Z_i^*\rangle(\infty) - \langle Z_i^*,Z_i^*\rangle(\phi^*(u)) \to_P 0 \quad \text{as } n \to \infty$$

and therefore (4.2.1) holds for all $t^* \in [0,\infty]$, recalling that $h_i = 0$ outside I by definition.

Now the function of t^* defined by the right hand side of (4.2.1) is continuous, so Theorem 2.4.1 can be applied to prove weak convergence of $\{Z_i^*: i = 1,\ldots,r\}$ on $(D(I^*))^r$, $(D([0,\phi^*(u)]))^r$ or $(D([0,\infty]))^r$ respectively according to whether Conditions I, I and II, or I, II and III have been imposed. Because we have weak convergence to a continuous limit the Skorohod construction can be applied (see Theorem 2.4.3 and the remarks following it) to replace $\to_{\mathcal{D}}$ with almost sure convergence in the supremum distance on a new probability space (except in the case of $D([0,\phi_-^*(u)))$, when we obtain almost sure convergence in the supremum distance on $[0,t^*]$ for each $t^* < \phi_-^*(u))$. By deleting all the intervals $[\phi_-^*(t),\phi^*(t))$ we obtain, on this new probability space, almost sure convergence in the supremum metric over all compact intervals of $\{Z_i: i = 1,\ldots,r\}$ to $\{Z_i^\infty: i = 1,\ldots,r\}$, where Z_i^∞ has all the required properties. Almost sure convergence implies convergence in distribution, so the theorem is proved. □

A few comments on the proof of this theorem are in order. When all the distribution functions concerned are continuous, the proof is a very direct application of Theorem 2.4.1, which is of course itself very much concerned with "the continuous case". In this part of the proof we only used Assumption 3.1.1. To accomodate jumps, we had to carry out a rather elaborate construction to bring us back to the continuous case, and needed Assumption 3.1.2 to do this. It is actually not very difficult to prove the above theorem in the "purely discrete case" - the random variables X_{ij} and $\tilde{X}_{ij}$ integer valued - rather more directly, using only Assumption 3.1.2 and the measurability requirements of Assumption 3.1.1. However it seems impossible to use Theorem 2.4.1 for the continuous part and the direct method for the discrete part in a mixed situation. A more elegant proof than the present one can probably be constructed by adapting the proof of LIPTSER & SHIRYAYEV's (1980) functional central limit theorem for semimartingales.

It should be noted that a version of Theorem 4.2.1 could have been proved with the interval I depending on i, $I = I_i$ say, giving weak convergence on $\prod_{i=1}^{r} D(I_i')$, where $I_i' = I_i$, $[0,u_i]$ or $[0,\infty]$ according to whether Conditions I, I and II, or I, II and III were supposed to hold for this i.

Our first application of Theorem 4.2.1 is to the product limit estimator. Take $r = 1$, drop the index i, and suppose that the distribution function F^n being estimated is fixed, say $F^n = F$ for all n.

THEOREM 4.2.2 *(Weak convergence of the product limit estimator).* *Suppose* $r = 1$ *and* $F^n = F$ *for all* n, *and suppose that* Y/n *converges uniformly on* $[0,\infty)$ *to a function* y *in probability as* $n \to \infty$. *Then*

$$n^{\frac{1}{2}}(\hat{F}-F) \to_{\mathcal{D}} (1-F)\, Z^{\infty} \quad \text{as } n \to \infty$$

on $D(\mathcal{I})$, *where* $\mathcal{I} = \{t\colon y(t) > 0\}$ *and* Z^{∞} *is a zero-mean Gaussian process with independent increments and variance function*

$$\operatorname{var}(Z^{\infty}(t)) = \int_0^t \frac{\chi_{[0,1)}(\Delta G)}{1 - \Delta G} \frac{dG}{y}$$

which may consistently be estimated by $n \int_0^t \frac{\chi_{\{\Delta N<Y\}}}{Y-\Delta N} \frac{dN}{Y}$; *if* $\hat{F}(t) < 1$ *we have*

$$\int_0^t \frac{\chi_{\{\Delta N<Y\}}}{Y-\Delta N} \frac{dN}{Y} = \frac{\hat{V}(t)}{(1-\hat{F}(t))^2}$$

(see (3.2.21)).

PROOF. As in Theorem 4.1.1 we use the representation (3.2.13) which we here rewrite as

$$n^{\frac{1}{2}}(\hat{F}-F) = (1-F) \int \frac{\chi_{[0,1)}(\Delta G)}{1-\Delta G} \frac{(1-\hat{F}_-)}{(1-F_-)} \frac{n^{\frac{1}{2}}J}{Y}\, dM$$

on $\{t\colon Y(t) > 0\}$. (If $F(t) = 1$ then on the event $\{Y(t) > 0\}$ we have, almost surely, $\Delta N(t) = Y(t)$ and hence $\hat{F}(t) = 1$.) Note that $y(t) > 0$ implies that $F_-(t) < 1$ and $G(t) < \infty$. We shall verify Condition I of Theorem 4.2.1, taking

$$H = \frac{\chi_{[0,1)}(\Delta G)}{(1-\Delta G)} \frac{1-\hat{F}_-}{1-F_-} \frac{n^{\frac{1}{2}}J}{Y}$$

(see 4.1.4) and taking $\mathcal{I}$ as defined in the theorem. The only nontrivial part of Condition I is Ib. By Theorem 4.1.1, we see that for all $t \in \mathcal{I}$,

$$\sup_{s\in[0,t]} |\hat{F}(s)-F(s)| \to_P 0 \quad \text{as } n \to \infty.$$

So for each $t \in \mathcal{I}$

$$\sup_{s\in[0,t]} \left| H^2(s)Y(s) - \left(\frac{\chi_{[0,1)}(\Delta G)}{1-\Delta G}\right)^2 y^{-1} \right| \to_P 0 \quad \text{as } n\to\infty.$$

Since $F^n = F$ for all n, we need only verify that the limit h of H^2Y is bounded on closed subintervals of I, which is clearly the case.

Thus Theorem 4.2.1 gives us weak convergence in $D(I)$ of $\int HdM$ to a process Z^∞, having the required properties, in particular such that

$$\mathrm{var}(Z^\infty(t)) = \int_0^t \left(\frac{\chi_{[0,1)}(\Delta G)}{1-\Delta G}\right)^2 y^{-1}(1-\Delta G)dG$$

$$= \int_0^t \frac{\chi_{[0,1)}(\Delta G)}{1-\Delta G}\frac{dG}{y}.$$

By Theorem 4.1.1 we also have

$$\sup_{s\in[0,t]} \left| \int_0^t \frac{dN}{Y} - G(t) \right| \to_P 0 \quad \text{as } n\to\infty$$

for each $t \in I$, so it is not difficult to show that

$$\sup_{s\in[0,t]} \left| \frac{n\hat{V}(s)}{(1-\hat{F}(s))^2} - \mathrm{var}(Z^\infty(s)) \right| \to_P 0 \quad \text{as } n\to\infty$$

for each $t \in I$. □

Theorem 4.2.1 of course also supplies us with a Skorohod construction in the uniform metric for $n^{\frac{1}{2}}(\hat{F}-F)$. We can take advantage of this fact when F is a discrete distribution, giving weights in I to points $t_1, t_2, \ldots$ only, in order to conclude that

$$\{n^{\frac{1}{2}}(\hat{F}(t_i) - F(t_i)): i = 1,2,\ldots\}$$

is asymptotically distributed as

$$\{(1-F(t_i))Z^\infty(t_i): i = 1,2,\ldots\}.$$

Theorem 4.2.2 can also be used to derive asymptotic confidence bands for F, conservative in the case that F has jumps. For let $t \in I$ be fixed, and note that the process $\{Z^\infty(s)/\sqrt{\mathrm{var}\, Z^\infty(t)}: s \in [0,t]\}$ has the same distribution as $\{B(\frac{\mathrm{var}(Z^\infty(s))}{\mathrm{var}(Z^\infty(t))}): s \in [0,t]\}$, where B is a standard Brownian motion on [0,1] with continuous paths: both these processes are Gaussian, with the same mean and covariance functions, and both have right continuous paths. So for all x,

$$P\left(\sup_{s\in[0,t]} \frac{|Z^{\infty}(s)|}{\sqrt{\mathrm{var}\, Z^{\infty}(t)}} \le x\right) \ge P\left(\sup_{s\in[0,1]} |B(s)| \le x\right),$$

and there is equality for all x if and only if the function $\mathrm{var}(Z^{\infty})$ is continuous on $[0,t]$. So for any $t \in \mathcal{I}$,

$$\liminf_{n\to\infty} P\left(\sup_{s\in[0,t]} \left|\frac{\hat{F}(s) - F(s)}{1 - \hat{F}(s)}\right| \le \frac{\hat{V}(t)^{\frac{1}{2}}}{1 - \hat{F}(t)} \cdot x\right)$$

$$\ge P\left(\sup_{s\in[0,1]} |B(s)| \le x\right)$$

$$= \sum_{k=-\infty}^{\infty} (-1)^k (\Phi((2k+1)x) - \Phi((2k-1)x)),$$

where Φ is the standard normal distribution (see FELLER (1971), page 343, BILLINGSLEY (1968) page 79, or RÉNYI (1963), though beware of misprints in the first two cases). RÉNYI (1953) gives a table of $P(\sup_{[0,1]} |B| \le y\sqrt{\frac{a}{1-a}})$ for various values of y and a, and WALSH (1962) page 334 reproduces the table with y denoted by A and a by A_1. Note that when there is no censoring,

$$\frac{\hat{V}(t)}{(1 - \hat{F}(t))^2} = n^{-1} \frac{\hat{F}(t)}{1 - F(t)},$$

and the above confidence bands reduce to those proposed in RÉNYI (1953).

HALL & WELLNER (1980) and GILLESPIE & FISHER (1979) propose other methods of basing confidence bands for F on the weak convergence of $n^{\frac{1}{2}}(\hat{F}-F)$ which may be superior in some respects; however our proposal seems to be the simplest to implement.

In Example 4.1.1, the conditions of Theorem 4.2.2 become $\frac{1}{n}\sum_{j=1}^{n} L_j^n(t) \to L(t)$ as $n \to \infty$ uniformly in $t \in [0,\infty)$, for some (sub)-distribution function L (see the remarks following (4.1.33)). In this case, $y = (1 - F_-)(1 - L_-)$. BRESLOW & CROWLEY (1974) prove Theorem 4.2.2 under the usual random censorship model with F and L continuous; MEIER (1975) sketches a proof under the fixed censorship model, also with F continuous. AALEN & JOHANSEN (1978) Theorem 4.6 give a result very close to our Theorem 4.2.2 in the case that F is continuous and has a hazard rate: they assume uniform integrability (in t and n) of n/Y and pointwise convergence in probability instead of uniform convergence in probability.

Otherwise their result is more general as it is concerned with estimation for a Markov chain.

Back in Example 4.1.1, we can in fact obtain a stronger result on weak convergence on $D[0,u]$, where $u = \sup\{t: y(t) \geq 0\}$:

THEOREM 4.2.3 *(Weak convergence of the product limit estimator on maximal closed interval under general random censorship).*
Suppose in the situation of Example 4.1.1 that $r = 1$, $F^n = F$ *for all* n, *and*

$$\frac{1}{n}\sum_{j=1}^{n} L_j^n(t) \to L(t) \quad \textit{uniformly on } [0,\infty) \quad \textit{as } n \to \infty$$

for some (sub)-distribution function L. *Define* $y = (1-F_-)(1-L_-)$, $\mathcal{I} = \{t: y(t) > 0\}$, *and* $u = \sup \mathcal{I}$. *Suppose that* $y(u) > 0$, *or alternatively that* $\Delta F(u) = 0$,

$$(4.2.2) \qquad \lim_{t\uparrow u} (F(u)-F(t))^2 \int_0^t ((1-F)(1-F_-)(1-L_-))^{-1}\, dF = 0,$$

and

$$(4.2.3) \qquad \lim_{t\uparrow u} \limsup_{n\to\infty} \int_{(t,u)} \frac{\chi_{[0,1)}(L_-^n)}{(1-L_-^n)} (1-\Delta G)\, dF = 0.$$

Then defining for each n $T = \sup\{t: Y(t) > 0\}$ *and* $F^T(t) = F(t\wedge T)$,

$$n^{\frac{1}{2}} \frac{1-F}{1-F^T}(\hat{F} - F^T) \to_{\mathcal{D}} \chi_{[0,u)} \cdot (1-F)\cdot Z^\infty + \chi_{\{u\}}\cdot U$$

as $n \to \infty$ *in* $D[0,u]$, *where* Z^∞ *is a zero-mean Gaussian process on* $\mathcal{I}$ *with independent increments and variance function*

$$\operatorname{var}(Z^\infty(t)) = \int_0^t ((1-F)(1-F_-)(1-L_-))^{-1}\, dF$$

and

$$U = \begin{cases} (1-F(u))Z^\infty(u) & \textit{if } y(u) > 0, \\ \lim_{t\, u} (1-F(t))Z^\infty(t) & \textit{if } y(u) = 0. \end{cases}$$

Since $\lim_{t\uparrow u} (1-F(t))Z^{\infty}(t)$ *almost surely exists, this does define a random element of* $D[0,u]$. *If* $y(u) = 0$ *and* $F(u-) = 1$, *then* $U = 0$.

If also F *is continuous and* $F(u) = 1$, *then*

$$\sup_{[0,u]} \left| n\hat{V} - (1-F)^2 \operatorname{var} Z^{\infty} \right| \to_P 0 \quad \text{as } n \to \infty.$$

PROOF. Note first that in the case $y(u) = 0$, (4.2.3) and (4.2.2) imply

$$(4.2.4) \qquad \int_I (1-L_-)^{-1}(1-\Delta G)\, dF < \infty$$

and

$$(4.2.5) \qquad \lim_{t\uparrow u} \limsup_{n\to\infty} (F(u)-F(t))^2 \cdot \int_0^t \chi_{[0,1)}(L_-^n) \cdot ((1-F)(1-F_-)(1-L_-^n))^{-1}\, dF = 0.$$

Next we shall show, using (4.2.2) and (4.2.4), that $\lim_{t\uparrow u} (1-F(t))Z^{\infty}(t)$ exists almost surely if $y(u) = 0$. Suppose $y(u) = 0$, and fix $s < u$ for the moment. On $[s,u)$, $(Z^{\infty}-Z^{\infty}(s))^2$ is a submartingale and by the well known Birnbaum-Marshall inequality (BIRNBAUM & MARSHALL (1961) Theorem 5.1),

$$P\left(\sup_{[s,u)} (1-F)^2(Z^{\infty}-Z^{\infty}(s))^2 \geq \varepsilon\right)$$

$$\leq \int_{[s,u)} \frac{(1-F)^2\, dF}{\varepsilon(1-F)(1-F_-)(1-L_-)} = \frac{1}{\varepsilon}\int_{[s,u)} (1-L_-)^{-1}(1-\Delta G)\, dF.$$

We have in fact used a slight sharpening of the inequality because BIRNBAUM & MARSHALL (1961) require that $(1-F)^2$ and $E((Z^{\infty}-Z^{\infty}(s))^2)$ have no jumps in common. However their proof is easily adapted to take care of this extension. Therefore

$$P\left(\sup_{[s,u)} ((1-F)\cdot Z^{\infty} - (1-F(s))Z^{\infty}(s))^2 \geq 2\varepsilon\right)$$

$$\leq \frac{1}{\varepsilon}\int_{[s,u)} (1-L_-)^{-1}(1-\Delta G)\, dF + P((F(u-)-F(s))^2(Z^{\infty}(s))^2 \geq \varepsilon)$$

$$\leq \frac{1}{\varepsilon}\int_{[s,u)} (1-L_-)^{-1}(1-\Delta G)\, dF + \frac{1}{\varepsilon}(F(u-)-F(s))^2 \operatorname{var}(Z^{\infty}(s)).$$

Let $\varepsilon_m > 0$ and $\delta_m > 0$, $m = 1,2,\ldots$, satisfy $\varepsilon_m \downarrow 0$ and $\sum_{m=1}^{\infty} \delta_m < \infty$. For each m by (4.2.2) and (4.2.4) and the fact that $y(u) = 0$, we can choose a

$\delta_m < u$ such that

$$P\left(\sup_{[s_m,u)} ((1-F)\cdot Z^\infty - (1-F(s_m))Z^\infty(s_m))^2 \geq 2\varepsilon_m\right) \leq \delta_m.$$

It is now easy to see by the Borel-Cantelli lemma that $\lim_{t\uparrow u} (1-F(t))Z^\infty(t)$ exists. Note that if $y(u) = 0$ and $F(u-) = 1$, then by (4.2.2), $(1-F(t))Z^\infty(t) \to_P 0$ as $t \uparrow u$ so in this case, $U = 0$.

Now we prove weak convergence of $n^{\frac{1}{2}}(1-F)(\hat{F}-F^T)/(1-F^T)$. Define for each n

$$Z = n^{\frac{1}{2}} \int \frac{1-\hat{F}_-}{1-F}\frac{J}{Y}\,dM = n^{\frac{1}{2}}\frac{(\hat{F}-F^T)}{1-F^T}$$

(replace t with $t\wedge T$ in (3.2.13)) so that

$$n^{\frac{1}{2}}\frac{1-F}{1-F^T}(\hat{F}-F^T) = (1-F)\cdot Z.$$

We already know by Theorem 4.2.2 that $(1-F)\cdot Z \to_{\mathcal{D}} (1-F)\cdot Z^\infty$ in $D[0,t]$ for each $t \in I$. So by BILLINGSLEY (1968) Theorem 4.2 it remains to show that if $u \notin I$, then

$$\lim_{t\uparrow u}\limsup_{n\to\infty} P\left(\sup_{s\in[t,u]} |(1-F(s))Z(s) - (1-F(t))Z(t)| > \varepsilon\right) = 0$$

for all $\varepsilon > 0$.

Suppose $y(u) = 0$, fix $t < u$ for the moment and note that

$$\sup_{[t,u]} |(1-F)\cdot Z - (1-F(t))Z(t)|$$

$$\leq \sup_{[t,u]} |(1-F)\cdot(Z-Z(t))| + (F(u)-F(t))|Z(t)|.$$

For each $t' \in (t,u]$ such that $F(t') < 1$, $Z-Z(t)$ is a square integrable martingale on $[t,t']$, and $(Z-Z(t))^2 - (\langle Z,Z\rangle - \langle Z,Z\rangle(t))$ is a martingale on $[t,t']$. Both processes are zero at time t and have paths of bounded variation. Also, for $s \in [t,t']$,

$$(1-F(s))^2(Z(s)-Z(t))^2 = \int_{(t,s]} (1-F)^2 d((Z-Z(t))^2) +$$

$$+ \int_{(t,s]} (Z_- - Z(t))^2 d((1-F)^2)$$

$$\leq \int_{(t,s]} (1-F)^2 d((Z-Z(t))^2).$$

Considered as a process, $(1-F)^2$ is predictable, so for any stopping time S taking values in [t,t'],

$$E((1-F(S))^2 (Z(S)-Z(t))^2)$$

$$\leq E\left(\int_{(t,S]} (1-F)^2 d((Z-Z(t))^2\right)$$

$$= E\left(\int_{(t,S]} (1-F)^2 d\langle Z,Z\rangle\right)$$

$$\leq E\left(\int_{(t,S]} (1-\hat{F}_-)^2 J \frac{n}{Y} (1-\Delta G) \frac{dF}{1-F_-}\right),$$

where the last inequality follows from (4.1.6). Theorem 2.4.2 therefore gives us

$$P\left(\sup_{[t,t']} |(1-F)(Z-Z(t))| > \varepsilon\right) \leq \frac{\delta}{\varepsilon^2} + P\left(\int_{[t,t']} \frac{(1-\hat{F}_-)^2}{(1-F_-)} J \frac{n}{Y} (1-\Delta G)\, dF > \delta\right).$$

If $F(u) < 1$ we can choose $t' = u$ in this relation; but otherwise letting $t' \uparrow u$ also shows that it is true with $t' = u$. By Theorem 3.2.1 and VAN ZUIJLEN (1978) Theorem 1.1 and Corollary 3.1,

$$P\left(\int_{(t,u]} \frac{(1-\hat{F}_-)^2}{(1-F_-)} J \frac{n}{Y} (1-\Delta G)\, dF \geq \beta^{-3} \int_{(t,u]} \frac{\chi_{[0,1)}(L_-^n)}{(1-L_-^n)} (1-\Delta G)\, dF\right) = o(1)$$

as $\beta \downarrow 0$ uniformly in n. Therefore by (4.2.3)

$$\lim_{t \uparrow u} \limsup_{n\to\infty} P\left(\sup_{[t,u]} |(1-F)(Z-Z(t))| > \varepsilon\right) = 0 \quad \text{for all } \varepsilon > 0.$$

It remains to show that

$$\lim_{t \uparrow u} \limsup_{n\to\infty} P((F(u)-F(t))|Z(t)| > \varepsilon) = 0 \quad \text{for all } \varepsilon > 0.$$

But because $Z^2 - \langle Z,Z\rangle$ is a martingale on [0,t], and $\langle Z,Z\rangle$ a nondecreasing predictable process, again by Theorem 2.4.2 we have

$$P((F(u)-F(t))|Z(t)| > \varepsilon)$$

$$\leq \frac{\delta}{\varepsilon^2} + P\left((F(u)-F(t))^2 \int_0^t \frac{(1-\hat{F}_-)^2}{(1-F)^2} J \frac{n}{Y}(1-\Delta G)dG > \delta\right)$$

By Theorem 3.2.1 and VAN ZUIJLEN (1978) Theorem 1.1 and Corollary 3.1

$$P\left(\int_0^t \frac{(1-\hat{F}_-)^2}{(1-F)^2} J \frac{n}{Y}(1-\Delta G)dG \geq \beta^{-3} \int_0^t \frac{\chi_{[0,1)}(L_-^n)}{(1-F)(1-F_-)(1-L_-^n)} dF\right) = o(1)$$

as $\beta \downarrow 0$ uniformly in n; and hence (4.2.5) yields the required result.

Next we consider the variance estimator $n\hat{V}$, supposing that F is continuous. If $y(u) > 0$ there is nothing to prove. So we suppose $y(u) = 0$; because $F(u) = 1$ this implies that $(1-F(t))^2 \operatorname{var} Z^\infty(t) \to 0$ as $t\uparrow u$. In view of Theorem 4.2.2 and the continuity of F, we only have to show that

$$\lim_{t\uparrow u} \limsup_{n\to\infty} P\left(\sup_{s\in[t,u]} (1-\hat{F}(s))^2 \int_0^s n \frac{\chi_{\{Y>1\}}}{Y-1} \frac{dN}{Y} > \varepsilon\right) = 0$$

for all $\varepsilon > 0$. Now by Theorem 3.2.1, it suffices to prove this with $1-\hat{F}(s)$ replaced by $1-F(s)$. Note also that because $\chi_{\{Y>1\}}/(Y\cdot(Y-1))$ is predictable and bounded

$$E\int n \frac{\chi_{\{Y>1\}}}{Y-1} \frac{dN}{Y} = E\int n \frac{\chi_{\{Y>1\}}}{Y-1} dG.$$

By the Birnbaum-Marshall inequality and the above remarks,

$$P\left(\sup_{s\in[t,u]} (1-F(s))^2 \int_0^s n \frac{\chi_{\{Y>1\}}}{(Y-1)Y} dN > \varepsilon\right)$$

$$\leq \frac{(1-F(t))^2}{\varepsilon} E\int_0^t n \frac{\chi_{\{Y>1\}}}{(Y-1)} dG + \int_{(t,u]} \frac{(1-F)^2}{\varepsilon} E\left(n \frac{\chi_{\{Y>1\}}}{Y-1}\right) dG.$$

Now

$$E\left(n \frac{\chi_{\{Y>1\}}}{(Y-1)}\right) \leq 3E\left(\frac{n+1}{(Y+1)}\right)\chi_{[0,1)}(L_-^n) \leq \frac{3\chi_{[0,1)}(L_-^n)}{(1-F_-)(1-L_-^n)},$$

where the final inequality holds by HOEFFDING (1956) Theorem 3. Relations (4.2.3) and (4.2.5) now yield the required result. □

Let us discuss some of the relationships between Conditions (4.2.2) to (4.2.4). If $L^n = L$ for all n and $y(u) = 0$, then (4.2.3) and (4.2.4) are equivalent.

Consider now the case in which $F(u) = F(u-) = 1$. We can write

$$(1-F)^2 \int ((1-F)(1-F_-)(1-L_-))^{-1}dF$$
$$= \int (1-L_-)^{-1}(1-\Delta G)\,dF + \int \left(\int ((1-F)(1-F_-)(1-L_-))^{-1}\,dF\right)d((1-F)^2),$$

where the first term on the right hand side is nondecreasing and the second nonincreasing and both are zero at time zero. So in this situation, (4.2.4) implies that the limit in (4.2.2) exists, though not necessarily that it equals zero.

Finally, suppose that F is continuous and $F(u) = 1$. If $(1-L) \geq c(1-F)^{\alpha}$ for some $\alpha < 1$ and $c > 0$, then (4.2.2) and (4.2.4) both hold; if on the other hand $(1-L) \leq c(1-F)$ for some $c > 0$ then (4.2.2) and (4.2.4) both fail.

Theorem 4.2.3 gives a positive answer to a conjecture of HALL & WELLNER (1980), so their paper now also provides a method for constructing confidence bands for F on $[0,u]$ instead of on $[0,t]$ for some $t < u$. Several authors (e.g. EFRON (1967), HOLLANDER & PROSCHAN (1979)) make use of weak convergence on $[0,u]$ when in fact the literature only provides weak convergence on $[0,u)$. The proof of Theorem 4.2.3 can be adapted to solve a long outstanding problem concerning the product limit estimator: how to use it to estimate mean lifetime when no $t < \infty$ exists such that $F(t) = 1$. We present a discussion of this problem and some preliminary results in Appendix 5. Of course in the bounded case just mentioned Theorem 4.2.3 can be applied directly.

4.3. Weak convergence: test statistics of the class K vgl. 4.2.1

Taking $r = 2$ and $H_i = K/Y_i$, $i = 1$ and 2, in Theorem 4.2.1 will give conditions for asymptotic normality under the null hypothesis of $W(\infty)$ (and more generally also of $W(T)$ for a possibly random time instant T); for under the null hypothesis we have

$$W = \int K\frac{dN_1}{Y_1} - \int K\frac{dN_2}{Y_2} = \int \frac{K}{Y_1}\,dM_1 - \int \frac{K}{Y_2}\,dM_2. \tag{4.3.1}$$

More details are given in Corollaries 4.3.1 and 4.3.2. However we must also prove consistency of the null hypothesis variance estimators $V_1(\infty)$ and $V_2(\infty)$. The next result establishes consistency under only slightly stronger conditions than those of Theorem 4.2.1. In it we also consider contiguous alternatives, so that the result can be used in Chapter 5 too. Note that Conditions (4.3.3) to (4.3.5) needed for consistency of $V_2(\infty)$ are empty under the null hypothesis.

LEMMA 4.3.1. *Consider the situation of Theorem* 4.2.1, *taking* $r = 2$ *and* $H_i = K/Y_i$, $i = 1,2$. *Suppose that Condition* I *holds, with the functions* h_i *left continuous with right hand limits and of bounded variation on closed subintervals of* $\mathcal{I}$ *even if* F_i^n *does not depend on* n. *Suppose that the limiting distribution functions* F_1 *and* F_2 *are equal,* $F_1 = F_2 = F$ *say. Then with* $\ell = 1$

$$(4.3.2) \qquad \sup_{s\in[0,t]} \left|V_\ell(s) - \sum_{i=1}^{2} \int_0^s h_i(1-\Delta G)dG\right| \to_P 0 \quad \text{as } n \to \infty$$

for each $t \in \mathcal{I}$. *If Condition* II *holds, we also have* (4.3.2) *with* $t = u$; *and with the further addition of Condition* III, (4.3.2) *holds with* $t = \infty$.

The same statement holds with $\ell = 2$ *if the following three conditions (for* $i = 1$ *and* 2) *are added to Conditions* I, II *and* III *respectively:*

$$(4.3.3) \qquad \int_0^t |dG_i^n - dG| \to 0 \quad \text{as } n \to \infty \text{ for all } t \in \mathcal{I};$$

(4.3.4) *If* $u \notin \mathcal{I}$,

$$\lim_{t\uparrow u} \limsup_{n\to\infty} \sup_{s\in(t,u]} \left|\frac{dG_i^n}{dG_{i'}^n}(s)\right| < \infty \qquad i' \neq i;$$

(4.3.5) *If* $u < \infty$,

$$\limsup_{n\to\infty} \sup_{s\in(u,\infty)} \left|\frac{dG_i^n}{dG_{i'}^n}(s)\right| < \infty \qquad i' \neq i.$$

PROOF. From (4.1.20) and (4.1.21) we see that

$$V_1 = \sum_{i=1}^{2} \int H_i^2\, Y_i\left(1 - \frac{\Delta N_i - 1}{Y_i - 1}\right)\frac{dN_i}{Y_i}$$

and

$$V_2 = \sum_{i=1}^{2} \int H_i^2\, Y_i\left(1 - \frac{\Delta N_1 + \Delta N_2 - 1}{Y_1 + Y_2 - 1}\right)\frac{d(N_1+N_2)}{Y_1+Y_2}.$$

So under Condition I with the extra conditions on h_i, it is easy to see that (4.3.2) holds for all $t \in \mathcal{I}$ if (for $\ell = 1$)

$$\sup_{s\in[0,t]} \left|\int_0^s \frac{dN_i}{Y_i} - G(s)\right| \to_P 0$$

and if (for $\ell = 2$)

$$\sup_{s\in[0,t]} \left|\int_0^s \frac{d(N_1+N_2)}{Y_1+Y_2} - G(s)\right| \to_P 0$$

as $n \to \infty$ for each $t \in \mathcal{I}$ and each $i = 1,2$. The first relation follows immediately from Theorem 4.1.1, while the second relation follows by writing, on

$\{s\colon Y_1(s)\wedge Y_2(s) > 0\}$

$$(4.3.6)\qquad \int_0^s \frac{d(N_1+N_2)}{Y_1+Y_2} - G = \int_0^s \frac{Y_1}{Y_1+Y_2}\left(\frac{dN_1}{Y_1} - dG_1^n\right) + \int_0^s \frac{Y_2}{Y_1+Y_2}\left(\frac{dN_2}{Y_2} - dG_2^n\right)$$
$$+ \int_0^s \frac{Y_1}{Y_1+Y_2}(dG_1^n - dG) + \int_0^s \frac{Y_2}{Y_1+Y_2}(dG_2^n - dG).$$

Using Theorem 2.4.2 in the same way as was done in Theorem 4.1.1 to prove consistency of $\int dN_i/Y_i$ as an estimator of G_i^n, we find for any i and any fixed $t \in I$ that

$$P\left(\sup_{s\in[0,t]} \left|\int_0^s \frac{Y_i}{Y_1+Y_2}\left(\frac{dN_i}{Y_i} - J_i dG_i^n\right)\right|^2 \geq \varepsilon\right)$$
$$\leq \frac{\eta}{\varepsilon} + P\left(\int_0^t \left(\frac{Y_i}{Y_1+Y_2}\right)^2 \frac{J_i}{Y_i}(1-\Delta G_i^n)dG_i^n > \eta\right)$$
$$\leq \frac{\eta}{\varepsilon} + P\left(\frac{G_i^n(t)}{Y_i(t)} > \eta\right)$$

and so the first two terms on the right hand side of (4.3.6) converge uniformly in probability to zero on each closed subinterval of I. The same holds for the last two terms by Assumption (4.3.3).

Suppose next that $u \notin I$ and that Condition II holds. For any $s \leq t \leq u$,

$$V_1(t) - V_1(s) \leq \sum_{i=1}^2 \int_{(s,t]} H_i^2 Y_i \frac{dN_i}{Y_i}$$

while as t varies,

$$\sum_{i=1}^2 \int_{(s,t]} H_i^2 Y_i \frac{dN_i}{Y_i} - \sum_{i=1}^2 \int_{(s,t]} H_i^2 Y_i dG_i^n$$

is a martingale on [s,u], zero at time s. By Theorem 2.4.2 therefore, for all $\varepsilon > 0$ and $\eta > 0$,

$$P(V_1(u) - V_1(s) > \varepsilon) \leq \frac{\eta}{\varepsilon} + P\left(\sum_{i=1}^2 \int_{(s,u]} H_i^2 Y_i dG_i^n > \eta\right).$$

So by Condition II,

$$\lim_{t\uparrow u} \limsup_{n\to\infty} P\left(\sup_{s\in(t,u]} |V_1(u) - V_1(s)| > \varepsilon\right) = 0$$

for all $\varepsilon > 0$. Using BILLINGSLEY (1968) Theorem 4.2 as usual and the fact that $\int_0^u h_i(1-\Delta G_i)dG_i$ is finite shows that (4.3.2) holds with $t = u$ and $\ell = 1$.

Adding Condition III, this argument may be extended to all $t \in [0,\infty)$, still with $\ell = 1$.

For $\ell = 2$ we note that for any $s \le t \le u$,

$$V_2(t) - V_2(s) \le \sum_{i=1}^{2} \int_{(s,t]} H_i^2 Y_i \frac{d(N_1+N_2)}{Y_1+Y_2}$$

while for each i and i', as t varies,

$$\int_{(s,t]} H_i^2 Y_i \frac{dN_{i'}}{Y_1+Y_2} - \int_{(s,t]} H_i^2 Y_i Y_{i'} \frac{dG_{i'}^n}{Y_1+Y_2}$$

is a martingale on $[s,u]$, zero at time s. So by Theorem 2.4.2, for each $\varepsilon > 0$ and $\eta > 0$,

$$P(V_2(u) - V_2(s) > \varepsilon) \le \frac{\eta}{\varepsilon} + \sum_{i=1}^{2} P\Big(\sum_{i'=1}^{2} \int_{(s,u]} H_i^2 Y_i dG_{i'}^n > \frac{\eta}{2}\Big)$$

$$\le \frac{\eta}{\varepsilon} + \sum_{i=1}^{2} P\Big(\int_{(s,u]} H_i^2 Y_i dG_i^n > \frac{\eta}{2} \,/\, (1+c)\Big)$$

for s sufficiently close to u, and n sufficiently large, where $c < \infty$ is some constant greater than the left hand side of (4.3.4). Using Condition II again gives us the required result for $t = u$ and $\ell = 2$.

Finally using (4.3.5) and Condition III in the same way for the case $t = \infty$ and $\ell = 2$ completes the proof. □

We can now give conditions for asymptotic normality of a test statistic of the class K (see page 55) in terms of the conditions I, II and III which were listed at the beginning of Section 4.2:

COROLLARY 4.3.1. *For each* n *let* $K \in K$ *be a random weight function generating test statistics* $W(\infty)/\sqrt{V_\ell(\infty)}$ *and more generally* $W(t)/\sqrt{V_\ell(t)}$ *for each* $t \in (0,\infty]$, $\ell = 1,2$. *Define* $H_i = K/Y_i$, $i = 1$ *and* 2, *and let* I *be an interval* $[0,u)$ *or* $[0,u]$ *for some* $u \in (0,\infty]$. *Then under the null hypothesis* $F_1^n = F_2^n = F$ *for all* n, *we have*

$$W(t) \to_{\mathcal{D}} N(0,\sigma^2(t))$$

and

$$V_\ell(t) \to_P \sigma^2(t) = \sum_{i=1}^{2} \int_0^t h_i(1-\Delta G)dG \qquad \ell = 1 \text{ and } 2$$

for each $t \in I$, $[0,u]$ *or* $[0,\infty]$ *according to whether Conditions* I, I *and* II *or* I, II *and* III *are satisfied. (Condition* I *must be satisfied with the extra conditions on* h_i *even though* F_i^n *does not depend on* n.*) Note that* h_i *satisfies*

$$H_i^2(t)Y_i(t) = K^2(t)/Y_i(t) \to_P h_i(t) \qquad t \in I$$

$$h_i(t) = 0 \qquad\qquad t \notin I.$$

Sometimes we shall be interested in the test statistic $W(T^n)$ for some random time T^n defined for each $n = 1,2,\ldots$ (cf. the discussion at the end of Section 4.1 on the test statistic of EFRON and Example 3.1.2, Type II censorship):

<u>COROLLARY 4.3.2.</u> *Consider the situation of Corollary* 4.3.1. *Let* T^n *be a random time instant such that* $T_n \to_P t_0$ *as* $n \to \infty$*; if* t_0 *is a jump point of* $\sigma^2(t)$ *(defined in Corollary* 4.3.1*) suppose that either*

$$P(T^n \in [t_0,t_0+\varepsilon)) \to 1 \quad \textit{as } n \to \infty \textit{ for all } \varepsilon > 0$$

or

$$P(T^n \in (t_0-\varepsilon,t_0)) \to 1 \quad \textit{as } n \to \infty \textit{ for all } \varepsilon > 0.$$

If Condition I *holds (with the extra conditions on* h_i*) and* $P(T^n \in I) \to 1$ *as* $n \to \infty$*, then*

$$W(T^n) \to_{\mathcal{D}} N(0,\sigma^2)$$

and

$$V_\ell(T^n) \to_P \sigma^2$$

where $\sigma^2 = \sigma^2(t_0)$ *unless* T^n *approaches* t_0 *from below, when* $\sigma^2 = \sigma^2(t_0-)$. *If* $P(T^n \in [0,u]) \to 1$, *but Conditions* I *and* II *hold, the same conclusion is valid; the conclusion remains true if* t_0 *is arbitrary but Conditions* I *to* III *hold.*

Let us consider the special case of the test statistics of GEHAN, EFRON and COX, for which we have (cf. 4.1.22) to (4.1.24)):

$$H_i^2 Y_i = \frac{K^2}{Y_i} = \begin{cases} \dfrac{n_{i'}}{n_1+n_2}\,\dfrac{n_i}{Y_i}\left(\dfrac{Y_1}{n_1}\right)^2\left(\dfrac{Y_2}{n_2}\right)^2 & \text{(GEHAN)} \\ \dfrac{n_{i'}}{n_1+n_2}\,\dfrac{n_i}{Y_i}(1-\hat{F}_{1-})^2(1-\hat{F}_{2-})^2 J_1 J_2 & \text{(EFRON)} \\ \dfrac{n_{i'}}{n_1+n_2}\,\dfrac{n_i}{Y_i}\left(\dfrac{Y_1}{n_1}\right)^2\left(\dfrac{Y_2}{n_2}\right)^2\left(\dfrac{n_1+n_2}{Y_1+Y_2}\right)^2 & \text{(COX)} \end{cases}$$

for $i = 1,2$, and $i' \neq i$. Suppose that functions y_1 and y_2 exist such that as $n \to \infty$

(4.3.7) $$\sup_{t\in[0,\infty)} \left| \frac{Y_i(t)}{n_i} - y_i(t) \right| \to_P 0 \qquad i = 1 \text{ and } 2$$

and suppose also that

(4.3.8) $$n_1 \wedge n_2 \to \infty, \qquad \frac{n_i}{n_1+n_2} \to \rho_i \in [0,1] \qquad i = 1 \text{ and } 2.$$

Recall from Section 4.1 that the functions y_i are of necessity left continuous, nonincreasing, take values in $[0,1]$, and are such that $y_i(1-F_-)^{-1}$ is nonincreasing. A sufficient condition for (4.3.7) to hold in Example 4.1.1 is that the average censoring distribution for each sample converges uniformly to some distribution, i.e.

(4.3.9) $$L_i^n = \frac{1}{n_i}\sum_{j=1}^{n_i} L_{ij}^n \to L_i \qquad \text{as } n \to \infty \text{ for each } i$$

uniformly on $[0,\infty)$ for some (sub)-distribution functions L_1 and L_2. In this case $y_i(1-F_-)^{-1} = (1-L_{i-})$; even when we are not in the situation of Example 4.1.1 we shall interpret $y_i(1-F_-)^{-1}$ as the "limiting average censoring distribution" for sample i.

Let us define

(4.3.10) $$\mathcal{I} = \{t\colon y_1(t) \wedge y_2(t) > 0\}.$$

Since $y_i(t) > 0$ implies that $1-F(t-) > 0$, G is finite on $\mathcal{I}$. It is now easy to see, using Theorem 4.1.1 for the test statistic of EFRON, that Condition I holds with this choice of $\mathcal{I}$ for each of the three test statistics, if we take

(4.3.11) $$h_i = \frac{\rho_{i'}}{y_i}k^2 \qquad i' \neq i$$

and hence (see Corollary 4.3.1)

$$(4.3.12) \qquad \sigma^2(t) = \int_0^t \frac{\rho_1 y_1 + \rho_2 y_2}{y_1 y_2} k^2(1-\Delta G)\,dG,$$

where

$$(4.3.13) \qquad k_G = y_1 y_2$$

$$(4.3.14) \qquad k_E = (1-F_-)^2 \chi_I$$

and

$$(4.3.15) \qquad k_C = \frac{y_1 y_2}{\rho_1 y_1 + \rho_2 y_2}.$$

In each case, k is the limit in probability of $\sqrt{\frac{n_1+n_2}{n_1 n_2}}\, K$.

The situation as regards conditions II and III is different for each test statistic. It will turn out that (4.3.7) and (4.3.8) are sufficient and almost sufficient in the case of the test statistic of GEHAN and COX respectively: to illustrate the "almost" we give a counterexample in which $W_C(\infty)$ is not asymptotically normal though (4.3.7) and (4.3.8) hold. We shall give conditions in the situation of Example 4.1.1 for II and III to hold for the test statistic of EFRON. These conditions seem close to being necessary for asymptotic normality of $W_E(\infty)$. Note that Condition III is often trivially true; e.g. if $F(u) = F(\infty)$ or if $P(Y_1(u+) \wedge Y_2(u+) = 0) \to 1$ as $n \to \infty$. In Example 4.1.1 the latter holds if $L_1^n(u) = 1$ or $L_2^n(u) = 1$ for all n.

First we give a useful lemma:

LEMMA 4.3.2. *Under the null-hypothesis, if (4.3.7) holds, then*

$$(4.3.16) \qquad \int_0^\infty y_i\,dG < \infty$$

and

$$(4.3.17) \qquad \sup_{t\in[0,\infty)} \left| \int_0^t \frac{Y_i}{n_i}\,dG - \int_0^t y_i\,dG \right| \to_P 0 \quad \textit{as } n \to \infty.$$

PROOF. (4.3.16) follows immediately from the fact that $y_i \le (1-F_-)$. Clearly (4.3.17) holds if $[0,\infty)$ is replaced by $[0,s]$ for any s such that $G(s) < \infty$. Define $\tau = \sup\{t: F(t) < 1\}$ and suppose $G(\tau) = \infty$. Then Y_i is almost surely zero on (τ,∞) for each n, and y_i is zero on (τ,∞). Also $\Delta F(\tau) = 0$ so that

$$E\left(\int_t^\tau \frac{Y_i}{n_i}\,dG\right) = E\left(\int_t^\tau \frac{dN_i}{n_i}\right) \le F(\tau) - F(t) \downarrow 0 \quad \text{as } t \uparrow \tau$$

uniformly in n. So (4.3.17) holds in the case $G(\tau) = \infty$ too by the usual arguments. □

PROPOSITION 4.3.1 *(Asymptotic normality under the null-hypothesis of the test statistic of GEHAN).*
Suppose that (4.3.7) *and* (4.3.8) *hold. Then with* I *defined by* (4.3.10) *and* h_i *by* (4.3.11) *and* (4.3.13)*, Conditions* I *to* III *hold under the null-hypothesis for the test statistic of GEHAN.*

PROOF. Condition I has been already verified, and Condition IIa follows by

$$\int h_i(1 - \Delta G)dG \le \int y_i dG < \infty \quad \text{(see (4.3.16))}.$$

For Conditions IIb and III, note that $H_{i'}^2 Y_{i'} \le Y_i/n_i$ for each i and i'. If $u = \sup I$ and i are such that $y_i(u) = 0$, then by Lemma 4.3.2

$$\lim_{t\uparrow u} \limsup_{n\to\infty} P\left(\int_{(t,u]} \frac{Y_i}{n_i}\,dG > \varepsilon\right) = 0$$

for all $\varepsilon > 0$, while if $u < \infty$ and y_i is zero on (u,∞), again by Lemma 4.3.2

$$\int_{(u,\infty)} \frac{Y_i}{n_i}\,dG \to_P 0. \qquad \square$$

PROPOSITION 4.3.2 *(Asymptotic normality under the null-hypothesis of the test statistic of EFRON).*
Suppose that (4.3.7) *and* (4.3.8) *hold and let* I *be defined by* (4.3.10) *and* h_i *by* (4.3.11) *and* (4.3.14)*. Then under the null-hypothesis Condition* I *holds for the test statistic of EFRON. In Example* 4.1.1, *under* (4.3.8) *and* (4.3.9)*, Condition* II *holds if for each* i

$$\text{(4.3.18)} \qquad \lim_{t\uparrow u} \limsup_{n\to\infty} \frac{n_{i'}}{n_1+n_2} \int_{(t,u]} \frac{\{\prod_{j=1}^2 \chi_{[0,1)}(L_{j-}^n)\}(1-F_-)^2}{(1 - L_{i-}^n)}\,dF = 0 \quad i' \ne i$$

so that in particular

$$\text{(4.3.19)} \qquad \rho_{i'} \int_I \frac{(1-F_-)(1-F)}{(1 - L_{i-})}\,dF < \infty;$$

Condition III *holds if for each* i

$$\text{(4.3.20)} \qquad \lim_{n\to\infty} \frac{n_{i'}}{n_1+n_2} \int_{(u,\infty)} \frac{\{\prod_{j=1}^2 \chi_{[0,1)}(L_{j-}^n)\}(1-F_-)^2}{(1 - L_{i-}^n)}\,dF = 0.$$

PROOF. Condition I has already been dealt with. So consider the situation of Example 4.1.1 with (4.3.9) holding.

Condition (4.3.19) is precisely IIa. For

$$\int_I h_i(1-\Delta G)\,dG = \rho_{i'} \int_I \frac{(1-F_-)}{Y_i}\frac{1-F}{1-F_-}\frac{dF}{1-F_-}$$

$$= \rho_{i'} \int_I \frac{(1-F_-)(1-F)}{(1-L_{i-})}\,dF.$$

Recalling that

$$H_i^2 Y_i = \frac{n_{i'}}{n_1+n_2} J_1 J_2 \frac{n_i}{Y_i}(1-\hat{F}_{1-})^2(1-\hat{F}_{2-})^2,$$

under Example 4.1.1 we obtain by Theorem 3.2.1 and VAN ZUIJLEN (1978) Theorem 1.1 and Corollary 3.1

$$P\left(H_i^2 Y_i \le \frac{n_{i'}}{n_1+n_2}\beta^{-5} J_1 J_2 \frac{(1-F_-)^4}{(1-F_-)(1-L_{i-}^n)} \text{ on } [0,\infty)\right) = 1 - \mathcal{O}(1)$$

as $\beta \downarrow 0$ uniformly in n. Conditions IIb and III now follow immediately from (4.3.18) and (4.3.20) respectively. □

Note that we only used Example 4.1.1 to supply a uniform bound for $P(Y_i/n_i \ge \beta(1-F_-)(1-L_{i-}^n)$ on $\{t: J_i(t) > 0\})$; so some extensions to other types of censoring can also be made. Note also that if $L_i^n = L_i$ for all n and $\rho_i \in (0,1)$ for each i, then (4.3.18) and (4.3.20) follow from the slightly strengthened form of (4.3.19):

(4.3.21) $$\int_I \frac{(1-F_-)^2}{(1-L_{i-})}\,dF < \infty, \qquad i = 1 \text{ and } 2.$$

If F is continuous and F(u) = 1, (4.3.18), (4.3.19) and (4.3.20) hold if for some $c > 0$ and $\alpha < 3$,

$$(1-L_i^n) > c(1-F)^\alpha \quad \text{for all i and n;}$$

(4.3.18) fails in this situation if for i = 1 or 2 and some $\alpha \ge 3$, $c > 0$, we have $\rho_{i'} \ne 0$ and $1-L_i < c(1-F)^\alpha$.

PROPOSITION 4.3.3 *(Asymptotic normality under the null-hypothesis of the test statistic of COX).*

Suppose that (4.3.7) *and* (4.3.8) *hold. Then with* I *defined by* (4.3.10) *and* h_i *by* (4.3.11) *and* (4.3.15), *Condition* I *holds for the test statistic of* COX. *If* $u \notin I$ *Condition* II *holds unless* $\Delta F(u) > 0$ *and for* $i = 1$ *or* 2, $\rho_i = 0$ *and* $y_i(u) > 0$. *If* $u < \infty$ *Condition* III *holds unless* $F(u) < F(\infty)$ *and for* $i = 1$ *or* 2, $\rho_i = 0$ *and* $y_i(u+) > 0$. *Condition* II *also holds if* $Y_1(u) \wedge Y_2(u) = 0$ *almost surely for all* n, *and Condition* III *if* $Y_1(u+) \wedge Y_2(u+) = 0$ *almost surely for all* n.

PROOF. Condition I has already been dealt with. Now

$$H_1^2 Y_1 + H_2^2 Y_2 = \frac{Y_1}{n_1}\frac{Y_2}{n_2}\frac{n_1+n_2}{Y_1+Y_2} \leq \frac{n_1+n_2}{n_{i'}}\frac{Y_i}{n_i} \qquad i' \neq i.$$

For $i = 1$ or 2, $\rho_{i'} > 0$ and by Lemma 4.3.2

$$\int_{(t,u)} \frac{n_1+n_2}{n_i'}\frac{Y_i}{n_i}\, dG \to_P \rho_{i'}^{-1} \int_{(t,u)} y_i dG.$$

So Condition II holds if $\Delta G(u) = 0$, or if $Y_1(u) \wedge Y_2(u) = 0$ almost surely for all n. If $\rho_i < 1$ and $y_i(u) = 0$,

$$\int_{\{u\}} \frac{n_1+n_2}{n_i'}\frac{Y_i}{n_i}\, dG \to_P 0,$$

so Condition II also holds if for $i = 1$ or 2, $\rho_i < 1$ and $y_i(u) = 0$. Similarly if for $i = 1$ or 2, $\rho_i < 1$ and $(y_i(u+) = 0$ or $F(\infty) = F(u))$,

$$\int_{(u,\infty)} \frac{n_1+n_2}{n_i'}\frac{Y_i}{n_i}\, dG \to_P \rho_{i'}^{-1} \int_{(u,\infty)} y_i dG = 0$$

and Condition III holds in this case too. Condition III holds trivially if $Y_1(u+) \wedge Y_2(u+) = 0$ almost surely for all n. Since $u \notin I$ implies $y_1(u) = 0$ or $y_2(u) = 0$ and $u < \infty$ implies $y_1(u+) = 0$ or $y_2(u+) = 0$, Conditions II and III can only fail in the situation described in the proposition. □

Let us discuss these results and compare them with what can be found in the literature. We shall neglect the fact that we consider variance estimators different from those of some authors, as was mentioned in Section 3.3. We therefore only consider the asymptotic normality of $W(\infty)$.

Our result on the test statistic of GEHAN is very general. GEHAN (1965) considers a permutation test based on $W_G(\infty)$, but BRESLOW (1970) shows how the theory of U-statistics can be applied under the usual model of random censorship (Example 4.1.1, with $L^n_{ij} = L_i$ for all i, j and n) to obtain asymptotic normality of $W_G(\infty)$, and sketches a modification to deal with fixed censorship (Example 3.1.4) under a condition equivalent to (4.3.9). He works with F continuous and $\rho_i \in (0,1)$.

Apart from the restriction to Example 4.1.1, our result on the test statistic of EFRON is also very satisfactory. Condition (4.3.19) seems to be a more or less necessary condition for asymptotic normality of $W_E(\infty)$.

EFRON (1967), working under the model of random censorship just mentioned and assuming F and L_i to be continuous and $\rho_i \in (0,1)$, also assumes that (4.3.19) holds in his sketch of a proof of asymptotic normality of $W_E(\infty)$. However his proof only establishes, in our terms, weak convergence of the process W_E on $D(\mathcal{I})$. So our results show that an extension to $D([0,\infty])$ is possible.

As we remarked in Section 4.1, it seems advisable to use $W_E(t)$ as test statistic for some t such that $y_1(t) > 0$ and $y_2(t) > 0$. EFRON (1967) makes this suggestion, but does not actually prove asymptotic normality in this case.

Finally we consider Proposition 4.3.3 on the test statistic of COX. CROWLEY & THOMAS (1975) prove asymptotic normality of $W_C(\infty)$ under the same random censorship model as above, assuming that F is continuous and $\rho_i \in (0,1)$. So our proposition generalizes this result.

We now show by a counterexample that Proposition 4.3.3 is not valid if only the Conditions (4.3.7) and (4.3.8) are imposed. More precisely, we show that in Example 4.1.1, $W_C(\infty)$ is not necessarily asymptotically normally distributed, even though (4.3.8) and (4.3.9) hold. We construct this counterexample by letting Condition II fail, which requires F to be discontinuous. However similar but more complicated counterexamples can be constructed with continuous F in which Condition III fails.

In Example 4.1.1, suppose that $u \notin \mathcal{I}$, $\rho_1 = 0$, $y_1(u) > 0$ and $\Delta F(u) > 0$. We must have $u < \infty$ and $y_2(u) = 0$. Since "$L^n_2 = L_2$ for all n" would imply that "$Y_2(u) = 0$ almost surely for each n", we must allow L^n_2 to vary with n (as in the model of fixed censorship). As we assume that (4.3.9) holds, we shall suppose that

$$L^n_2(u-) < 1 \text{ for all } n, \quad L^n_2(u-) \to L_2(u-) = 1 \quad \text{as } n \to \infty.$$

To avoid degeneracies, we strengthen our previous assumptions slightly to

$$L_1(u-) < 1, \quad 0 < F(u-) < F(u) < 1, \quad \text{and} \quad \rho_1 = 0.$$

Now suppose that for some $v_1 < u < v_2$, F is constant on $[v_1,u)$ and on $[u,v_2)$. Suppose also that $L_i(v_1) < 1$ for each i, and that $L_i^n(v_2-) = 1$ for each i and n. In this situation

$$W_C(\infty) = W_C(v_1) + \Delta W_C(u),$$

where under the null-hypothesis, by (4.1.3), (4.1.18), (4.1.19) and (4.1.24)

$$(4.3.23) \qquad \Delta W_C = \sqrt{\frac{n_1+n_2}{n_1n_2}}\left(\frac{Y_1Y_2}{Y_1+Y_2}\frac{\Delta M_1}{Y_1} - \frac{Y_1Y_2}{Y_1+Y_2}\frac{\Delta M_2}{Y_2}\right)$$

$$= Y_1^{\frac{1}{2}}\left(\frac{\Delta N_1}{Y_1} - \Delta G\right)\sqrt{\frac{n_1+n_2}{n_2}}\left(\frac{Y_1}{n_1}\right)^{\frac{1}{2}}\frac{Y_2}{n_1}\left(\frac{Y_1}{n_1}+\frac{Y_2}{n_1}\right)^{-1}J_1J_2$$

$$- Y_2^{\frac{1}{2}}\left(\frac{\Delta N_2}{Y_2} - \Delta G\right)\sqrt{\frac{n_1+n_2}{n_2}}\frac{Y_1}{n_1}\left(\frac{Y_2}{n_1}\right)^{\frac{1}{2}}\left(\frac{Y_1}{n_1}+\frac{Y_2}{n_1}\right)^{-1}J_1J_2.$$

We first show that if $\frac{Y_2(u)}{n_1} \to_P c \in [0,\infty]$ as $n \to \infty$, then $\Delta W_C(\infty)$ and $W_C(v_1)$ are asymptotically independent and

$$(4.3.24) \qquad \Delta W_C(u) \to_{\mathcal{D}} N\left(0, \Delta G(u)(1-\Delta G(u))\frac{y_1(u)\,c}{y_1(u)+c}\right).$$

(We already know that $W_C(v_1) \to_{\mathcal{D}} N(0,\sigma^2)$ for some $\sigma^2 > 0$.) Note that it is always possible to construct L_2^n such that $\frac{Y_2(u)}{n_1} \to_P c$ for a given c; we have

$$E\left(\frac{Y_2(u)}{n_1}\right) = \frac{n_2}{n_1}(1-F(u-))(1-L_2^n(u-))$$

and

$$\operatorname{var}\left(\frac{Y_2(u)}{n_1}\right) \le \frac{1}{n_1}E\left(\frac{Y_2(u)}{n_1}\right);$$

if $c \in (0,\infty)$ we can then define $L_2^n(u-)$ by

$$(1-F(u-))(1-L_2^n(u-)) = \frac{n_1}{n_2}c$$

for sufficiently large n; otherwise we define $L_2^n(u-)$ by

$$(1 - F(u-))(1 - L_2^n(u-)) = \frac{n_1}{n_2} c_n$$

for all n, where c_n is suitably chosen so that in particular $c_n \downarrow 0$ if $c = 0$ and $c_n \uparrow \infty$ if $c = \infty$.

Since $Y_1(u)/n_1 \to_P y_1(u) > 0$ and $\frac{n_1+n_2}{n_2} \to 1$ as $n \to \infty$, while $Y_i^{\frac{1}{2}}(\frac{\Delta N_i}{Y_i} - \Delta G)$ is bounded in probability as $n \to \infty$, the case $c = 0$ is immediate ($EY_i(\frac{\Delta N_i}{Y_i} - \Delta G)^2 = \Delta G(1 - \Delta G)EJ_i$ by Assumption 3.1.2). If $c > 0$, then $Y_1(u) \to_P \infty$ and $Y_2(u) \to_P \infty$ as $n \to \infty$ and it is now easy to show, using Assumption 3.1.2, that

$$W_C(v_1), \quad Y_1(u)^{\frac{1}{2}}\left(\frac{\Delta N_1(u)}{Y_1(u)} - \Delta G(u)\right) \quad \text{and} \quad Y_2(u)^{\frac{1}{2}}\left(\frac{\Delta N_2(u)}{Y_2(u)} - \Delta G(u)\right)$$

are asymptotically independently normally distributed with means zero and variances σ^2, $(1 - \Delta G(u))\Delta G(u)$, and $(1 - \Delta G(u))\Delta G(u)$ respectively. So (4.3.24) holds in this case too.

We now obtain our counterexample by constructing the L_2^n's so that $Y_2(u)/n_1$ converges in probability to different values of c along different subsequences; then $W_C(\infty)$ does not converge in distribution along the whole sequence.

Actually this is not a counterexample to asymptotic normality of $W_C(\infty)/\sqrt{V_{C\ell}(\infty)}$, $\ell = 1$ or 2; for provided $\sigma^2 > 0$, it is easy to see that along each subsequence for which $Y_2(u)/n_1 \to_P c$ for some c, $V_{C\ell}(\infty)$ converges in probability to the asymptotic variance of $W_C(\infty)$, and hence

$$(4.3.25) \qquad W_C(\infty)/\sqrt{V_{C\ell}(\infty)} \to_{\mathcal{D}} N(0,1)$$

along this subsequence. From *any* subsequence a further subsequence can be extracted along which $Y_2(u)/n_1$ converges in probability and therefore (4.3.25) holds along the original sequence.

However, the example illustrates the complications that arise in the situation excluded in Proposition 4.2.3. Similar difficulties arise in proving consistency, which was why we assumed $\rho_i \in (0,1)$ in Section 4.1 for the test statistic of COX.

CHAPTER 5

EFFICIENCIES AND NEW TEST STATISTICS

5.1. Introduction; comparison of variance estimators

In this chapter we shall again be concerned with asymptotic results for the two-sample case, the basic notations and definitions having been summarized in Section 4.1 (see especially formulae (4.1.1) to (4.1.5) and (4.1.18) to (4.1.24)). In Section 5.2 we show how the methods of the previous chapter can be extended to prove asymptotic normality under a contiguous sequence of alternative hypotheses of test statistics of the class K. The limiting distribution has the same variance as under the null-hypothesis but a different expectation, from which Pitman asymptotic relative efficiencies can immediately be calculated and used to compare test statistics of the class. We shall of course pay special attention to the test statistics of GEHAN, EFRON and COX.

It should be recalled that COX derived his test statistic with the alternative hypothesis in mind

$$\frac{(1-\Delta G_2)}{(1-\Delta G_1)}\,\frac{dG_1}{dG_2} = \text{constant},$$

a so called "proportional odds" model. In the continuous case, this reduces to the alternative of "proportional hazards", also known as a "Lehmann alternative", dG_1/dG_2 = constant. It turns out that COX's test statistic is indeed the best of the class K for alternatives of proportional odds. This generalizes previous results concerning the usual model of random censorship (Example 4.1.1 with $L^n_{ij} = L_i$ for all i, j and n) and continuous F^n_1 and F^n_2.

On the other hand the test-statistics of GEHAN and EFRON seem to have no general optimality properties; their behaviour relative to the best test for a given type of alternative hypothesis depends on what we shall call the "limiting average censoring distributions" for each sample (in Example

4.1.1, these are the L_1 and L_2 defined by (4.3.9)).

In the case of random censorship and continuous F_i^n's just mentioned, it is known that the test statistic of COX is asymptotically most powerful against a contiguous proportional hazards alternative if and only if $L_1 = L_2$. We shall show that this result is much more generally true, and offer an intuitive explanation. We also suggest that any nonparametric type test can only be asymptotically most powerful against a particular contiguous alternative if $L_1 = L_2$, and suggest that even if $L_1 \neq L_2$ the best test of the class K for a particular alternative is in fact an optimal test in the wider class of nonparametric-type tests.

In Section 5.3 we concentrate on constructing tests which should be especially powerful against parametric alternatives which can be reduced to a location family after a suitable transformation, i.e.

$$F_i^n(x) = \Psi(g(x)+\theta_i^n) \qquad \theta_1^n \neq \theta_2^n,$$

where Ψ is a fixed continuous distribution function on $(-\infty,\infty)$, g is a fixed monotone transformation and θ_1^n and θ_2^n are real parameters. We determine the best test of the class K for given Ψ (we shall have to consider random weight functions which are not necessarily nonnegative). It turns out as expected that such a test is asymptotically most powerful if and only if the limiting average censoring distributions for the two samples are equal.

As an example, when Ψ is the standard normal distribution function and there is no censoring, this procedure supplies us with a new non-parametric test statistic, which is asymptotically uniformly most powerful and which unlike the test statistics of Fisher-Yates or Van der Waerden can be used with censored observations as well. We give conditions for asymptotic normality of this test statistic which cover the case of no censoring.

Le Cam's theory of contiguity is very useful in this section, allowing us to evaluate limiting distributions only under the null-hypothesis in order to determine efficiencies with respect to the likelihood-ratio test.

In Section 5.4 we pay attention to the question of how two-sample tests can be constructed which are consistent against a wider class of alternatives than those considered in Section 4.1. Since for a given random weight function K we can use W(s) as a test statistic for any value of s, it seems worth considering whether a test can be based on $\sup_{s\in[0,t]} |W(s)|$ for some chosen t. It turns out that such a test is consistent against the alternative $F_1 \neq F_2$ on $[0,t]$. One would expect to pay for this by a loss of power against an alternative to which $W(t)$ is suited. However we indicate that for an alternative of the ordered hazards type, and for small values of the size α of the tests, the two tests are asymptotically nearly equally powerful: the limit as size $\alpha \to 0$ of their Pitman asymptotic relative efficiency (which depends on α) equals 1.

All this time we have made no comparison of the two null hypothesis variance estimators $V_1(\infty)$ and $V_2(\infty)$ (see (4.1.20) and (4.1.21)) and unfortunately there are reasons for preferring either. Under the null hypothesis we would expect $V_2(\infty)$, which in effect combines the two samples in order to estimate G, to be a better estimator of the asymptotic variance of $W(\infty)$. However this same fact leads to extra difficulties and sometimes extra conditions in dealing with $V_2(\infty)$ both under contiguous and under fixed alternative hypotheses, and this suggests that its convergence in probability as $n \to \infty$ might be slower in such cases.

Under the null hypothesis or a contiguous sequence of alternatives, $V_1(\infty)$ and $V_2(\infty)$ generally both converge in probability to the asymptotic variance of $W(\infty)$. Under a fixed alternative they have different limits; and other things being equal one would prefer the variance estimator with the smaller limiting value.

Suppose then that $F_1^n = F_1$ and $F_2^n = F_2$ for all n, where $F_1 \neq F_2$. Suppose as usual that for each $i = 1,2$, $\frac{Y_i}{n_i}$ converges uniformly on $[0,\infty)$ to a function y_i as $n \to \infty$, in probability. Define

$$\mathcal{I} = \{t: y_1(t) \wedge y_2(t) > 0\};$$

we shall have $\mathcal{I} = [0,u]$ or $[0,u)$ for some $u \in (0,\infty]$ and G_1 and G_2 are finite on $\mathcal{I}$. Suppose also that

$$n_1 \wedge n_2 \to \infty, \quad \frac{n_i}{n_1+n_2} \to \rho_i \in [0,1] \quad \text{as } n \to \infty$$

and that for each $t \in \mathcal{I}$,

$$\sqrt{\frac{n_1+n_2}{n_1 n_2}}\, K$$

converges uniformly on $[0,t]$ to k as $n \to \infty$, in probability, where k is left continuous with right hand limits and k_+ of bounded variation on $[0,t]$. We define $k = 0$ outside $\mathcal{I}$. Writing

$$V_1 = \sum_{i=1}^{2} \frac{n_{i'}}{n_1+n_2} \int \left(\frac{n_1+n_2}{n_1 n_2} K^2\right)\cdot\left(\frac{n_i}{Y_i}\right)\cdot\left(1 - \frac{\Delta N_i - 1}{Y_i - 1}\right) \frac{dN_i}{Y_i} \qquad (i' \neq i)$$

and

$$V_2 = \sum_{i=1}^{2} \frac{n_i}{n_1+n_2} \int \left(\frac{n_1+n_2}{n_1 n_2} K^2\right)$$

$$\cdot \frac{n_{i'}}{Y_{i'}}\left(1 - \frac{Y_1}{Y_1+Y_2-1}\frac{\Delta N_1}{Y_1} - \frac{Y_2}{Y_1+Y_2-1}\frac{\Delta N_2 - 1}{Y_2}\right)\frac{dN_i}{Y_i} \qquad (i' \neq i)$$

(see (4.1.20) and (4.1.21)) it follows by (4.1.15) that in probability, V_1 and V_2 converge uniformly on $[0,t]$ to the functions

$$(5.1.1) \qquad \sum_{i=1}^{2} \rho_{i'} \int \frac{k^2}{y_i}(1 - \Delta G_i)\, dG_i \qquad (i' \neq i)$$

and

$$(5.1.2) \qquad \sum_{i=1}^{2} \rho_i \int \frac{k^2}{y_{i'}}\left(1 - \frac{\rho_1 y_1 \Delta G_1 + \rho_2 y_2 \Delta G_2}{\rho_1 y_1 + \rho_2 y_2}\right) dG_i \qquad (i' \neq i)$$

as $n \to \infty$, for each $t \in I$.

Under some further conditions (compare the use of Conditions II and III in the proof of Lemma 4.2.1) this also holds with the interval $[0,t]$ for $t \in I$ replaced with $[0,\infty]$. The interesting point however is that the two functions in (5.1.1) and (5.1.2) are not necessarily equal, and it is not true that one of them is always greater than or equal to the other. Thus a general choice between V_1 and V_2 cannot be based on these considerations either.

5.2. Efficiencies

In this section we apply Theorem 4.2.1 to the two-sample situation in which for each i

$$(5.2.1) \qquad F_i^n(t) \to F(t) \text{ uniformly in } t \in [0,\infty) \qquad \text{as } n \to \infty$$

for some fixed distribution function F, with respect to which F_i^n is absolutely continuous for each i and n. We suppose that this convergence is such that for some real valued functions γ_i,

$$(5.2.2) \qquad \sqrt{\frac{n_1 n_2}{n_1+n_2}}\left(\frac{dG_i^n}{dG}(t) - 1\right) \to \gamma_i(t) \qquad \text{as } n \to \infty$$

uniformly on each closed subinterval of $\{t\colon F(t-) < 1\}$, and we define

$$(5.2.3) \qquad \gamma = \gamma_1 - \gamma_2.$$

(In Section 5.3 we shall weaken these assumptions somewhat.) At the same time we suppose as in Section 4.3 (see (4.3.7) and (4.3.8)) that

(5.2.4) $$\frac{Y_i(t)}{n_i} \to y_i(t) \text{ uniformly on } [0,\infty) \text{ in probability}$$

and

(5.2.5) $$n_1 \wedge n_2 \to \infty, \quad \frac{n_i}{n_1+n_2} \to \rho_i \in [0,1]$$

for each i as $n \to \infty$. Define

(5.2.6) $$I = \{t\colon y_1(t) \wedge y_2(t) > 0\}, \qquad u = \sup I.$$

From the remarks preceding Theorem 4.1.3 on page 66, we recall that the functions y_i are such that $y_i(1-F_-)^{-1}$ has all the properties of 1 minus the left continuous version of a (sub)-distribution function: it is left continuous, nonincreasing, nonnegative, and takes the value 1 at time zero. In Example 4.1.1, if (4.3.9) holds, then $y_i(1-F_-)^{-1} = (1-L_{i-})$, i = 1 and 2, where L_i is the limiting average censoring distribution for sample i. However even when we are not in the situation of Example 4.1.1, we propose *defining* the *limiting average censoring distribution* L_i by $(1-L_{i-}) =$ $= y_i(1-F_-)^{-1}$.

Finally let $K \in K$ be a random weight function for each n, generating a sequence of test statistics $W(\infty)/\sqrt{V_\ell(\infty)}$ (cf. Section 4.1, especially (4.1.18) to (4.1.21)), such that

(5.2.7) $$\sqrt{\frac{n_1+n_2}{n_1 n_2}}\, K(t) \to k(t) \text{ uniformly on closed subintervals of } I$$

in probability as $n \to \infty$, where k is left continuous with right hand limits and k_+ of bounded variation on closed subintervals of I. Define k = 0 outside I. We call k a "limiting weight function".

As a consequence of (5.2.1) to (5.2.7), writing

(5.2.8) $$W = \int \frac{K}{Y_1}\, dM_1 - \int \frac{K}{Y_2}\, dM_2 + \int K\left(\frac{dG_1^n}{dG} - 1\right)dG - \int K\left(\frac{dG_2^n}{dG} - 1\right)dG$$

(here $M_i = N_i - \int Y_i dG_i^n$) and letting for i = 1 and 2

(5.2.9) $$H_i = K/Y_i,$$

then we have

$$H_i^2 Y_i = \frac{n_{i'}}{n_1+n_2} \frac{n_i}{Y_i} \frac{n_1+n_2}{n_1 n_2} K^2 \qquad (i' \neq i)$$

and Condition I of Section 4.2 holds with

$$(5.2.10) \qquad h_i = \frac{\rho_{i'}}{y_i} k^2,$$

so that

$$(5.2.11) \qquad h_1 + h_2 = \frac{\rho_1 y_1 + \rho_2 y_2}{y_1 y_2} k^2.$$

Note that condition (4.3.3) of Lemma 4.3.1 is a consequence of (5.2.2). If also

$$(5.2.12) \qquad \int_0^t |k\gamma_i| dG < \infty \quad \text{for all } t \in I \text{ and } i = 1,2,$$

then by (5.2.2) and (5.2.7),

$$(5.2.13) \qquad \sup_{s\in[0,t]} \left| \int_0^s K\left(\frac{dG_1^n}{dG} - 1\right) dG - \int_0^s k\gamma_i \, dG \right| \to_P 0$$

for all $t \in I$ and each $i = 1,2$. We can extend (5.2.13) to $t = u$ and then to $t = \infty$ in the usual way by making the extra assumptions

II* If $u \notin I$, then for $i = 1$ and 2

a) $\int_I |k\gamma_i| dG < \infty$

b) $\lim_{t\uparrow u} \limsup_{n\to\infty} P(\int_{[t,u]} |K| |dG_i^n - dG| > \varepsilon) = 0$ for all $\varepsilon > 0$

and

III* if $u < \infty$, then for $i = 1$ and 2

$\int_{(u,\infty)} |K| |dG_i^n - dG| \to_P 0$ as $n \to \infty$.

By Theorem 4.2.1, Lemma 4.3.1, and (5.2.13) we therefore have if (5.2.1) to (5.2.7) and (5.2.12) hold

$$(5.2.14) \qquad W(t) \to_{\mathcal{D}} N\left(\int_0^t k\gamma dG, \int_0^t \frac{\rho_1 y_1 + \rho_2 y_2}{y_1 y_2} k^2 (1 - \Delta G) dG\right)$$

for all $t \in I$ and $V_\ell(t)$ is a consistent estimator of the asymptotic variance in (5.2.14) for $\ell = 1$ and 2. If $u \notin I$ but Conditions II, II*, and (for the case $\ell = 2$) (4.3.4) hold, this is also true for $t = u$; and if

$u < \infty$ but III, III* and (for the case $\ell = 2$) (4.3.5) hold too, then it is true for all $t \in [0,\infty]$.

Suppose we are interested in some parametric family of distributions, and select a sequence $\{(F_1^n, F_2^n): n = 1,2,\ldots\}$ of pairs of distribution functions from this family such that (5.2.1) to (5.2.7) and (5.2.12) hold for certain functions k, y_1, y_2 and γ. Suppose that under the null-hypothesis sequence $F_1^n = F_2^n = F$ for all n, (5.2.1) to (5.2.7) and (5.2.12) hold with the same k, y_1 and y_2 but with $\gamma = 0$. Then under the appropriate set of conditions, the asymptotic relative efficiency (for this sequence of alternatives) of one test statistic $W(t)/\sqrt{V_\ell(t)}$ with respect to another is given by the ratio of their *efficacies*

$$(5.2.15) \qquad e(k,t) = \frac{\left(\int_0^t k\gamma dG\right)^2}{\left(\int_0^t \frac{\rho_1 y_1 + \rho_2 y_2}{y_1 y_2} k^2 (1-\Delta G) dG\right)},$$

the efficacy of such a test statistic depending on its limiting weight function k and the time instant t for given y_1, y_2, γ and G.

Recall from Chapter 4 that for the test statistics of GEHAN, EFRON and COX, (5.2.7) holds with

$$(5.2.16) \qquad k_G = y_1 y_2$$

$$(5.2.17) \qquad k_E = (1-F_-)^2 \chi_I$$

and

$$(5.2.18) \qquad k_C = \frac{y_1 y_2}{\rho_1 y_1 + \rho_2 y_2}.$$

It is a straightforward matter to extend Propositions 4.3.1 to 4.3.3 to cover the contiguous alternative hypothesis case. In particular Lemma 4.3.2 remains valid under (5.2.1). However we shall not go into these details here, nor discuss conditions for II* and III* to hold.

The following lemma establishes that

$$k \propto \frac{y_1 y_2}{\rho_1 y_1 + \rho_2 y_2} \frac{\gamma}{1-\Delta G} \quad \text{on } [0,t]$$

maximizes (5.2.15) over the function k. Note that with such a choice of k, the terms corresponding to asymptotic mean and variance in (5.2.15) are equal to one another and hence also to the efficacy itself.

LEMMA 5.2.1. *Let* $t \in (0,\infty]$ *be fixed and define*

$$\beta = \frac{\rho_1 y_1 + \rho_2 y_2}{y_1 y_2}(1 - \Delta G) \quad \text{on } [0,t].$$

Suppose

$$0 < \int_0^t \frac{\gamma^2}{\beta}\, dG < \infty.$$

Then if almost everywhere (dF) *on* $[0,t]$

$$k \propto \frac{\gamma}{\beta} \quad \text{where } \beta \neq 0,$$

k *maximizes* $e(k,t)$ *over all* k *such that*

$$0 < \int_0^t k^2 \beta dG < \infty.$$

PROOF. We can equivalently maximize $e(k,t)$ over all k such that the denominator in (5.2.15) $\int_0^t k^2\beta dG$ is fixed and equal to $\alpha > 0$. The theory of Lagrange multipliers then leads us to consider the problem of maximizing

$$\int_0^t k\gamma dG - \lambda\left(\int_0^t k^2 \beta dG - \alpha\right)$$

over all k, for some fixed λ. Bringing the integrands under a single integral sign and maximizing pointwise, assuming $\lambda > 0$ this problem has as solution

$$k = \frac{1}{2\lambda}\frac{\gamma}{\beta} \quad \text{where } \beta \neq 0.$$

By the assumptions $\gamma = 0$ where $\beta = 0$ almost everywhere (dF), so we can neglect the case $\beta = 0$. Since for a fixed $\lambda > 0$ we can choose $\alpha \neq 0$ such that $\int_0^t k^2\beta dG = \alpha$ with this choice of k, the same k is the solution of the constrained problem. □

Now y_1 and y_2 depend on the limiting average censoring distributions, which we may consider as arbitrary. So by Lemma 5.2.1, a test statistic in K with limiting weight function k can only be "optimal relative to γ" (in the sense of maximizing $e(k,t)$ for the appropriate t) if $k(\rho_1 y_1 + \rho_2 y_2)/(y_1 y_2)$ is proportional to $\gamma(1 - \Delta G)^{-1}$ and so, apart from a constant of proportionality which may depend on L_1 and L_2, only depends on F and γ. This shows that the test statistics of GEHAN and EFRON will only be optimal relative to γ when special relationships hold between γ, F, L_1 and L_2; i.e. under special limiting average censoring distributions. We shall come across some cases of this later. However the test statistic of COX is "optimal" if $\gamma(1-\Delta G)^{-1}$ is constant almost everywhere (-dG) except possibly where $\Delta G = 1$.

We shall show that this case arises if

$$(5.2.19)\qquad (1-\Delta G_i^n)^{-1}\, dG_i^n = \theta_i^n (1-\Delta G)^{-1}\, dG, \quad i = 1 \text{ and } 2,$$

(i.e. a proportional odds model) where

$$(5.2.20)\qquad \theta_1^n = 1 + c\sqrt{\frac{n_2}{n_1(n_1+n_2)}}$$

$$(5.2.21)\qquad \theta_2^n = 1 - c\sqrt{\frac{n_1}{n_2(n_1+n_2)}}$$

for some $c \neq 0$. Special cases are the geometric distribution and the Weibull distribution (with fixed shape but varying scale parameter); the latter including the exponential distribution. Under (5.2.19) to (5.2.21) we have, for i = 1 and 2,

$$(1-\Delta G)\,dG_i^n = \theta_i^n(1-\Delta G_i^n)\,dG$$

$$\Rightarrow\quad dG_i^n - dG = (\theta_i^n-1)\,dG - \theta_i^n\Delta G_i^n dG + \Delta G dG_i^n$$

$$= (\theta_i^n-1)\,dG - (\theta_i^n-1)\Delta G_i^n dG$$

$$\Rightarrow\quad \frac{dG_i^n}{dG} - 1 = (\theta_i^n-1)(1-\Delta G_i^n).$$

So as $n \to \infty$, $\sqrt{\frac{n_1 n_2}{n_1+n_2}}(\frac{dG_i^n}{dG} - 1)$ converges uniformly on $[0,t]$ to $\rho_2 c(1-\Delta G)$ or $-\rho_1 c(1-\Delta G)$ according to whether i = 1 or 2, if t satisfies $F(t-) < 1$. Thus (5.2.2) holds with

$$(5.2.22)\qquad \gamma = \gamma_1 - \gamma_2 = c(1-\Delta G).$$

In Figure 5.2.1 we plot $e(k,t)$ for $k = k_G$, k_E and k_C as functions of t in the case that

$$F(t) = 1 - e^{-t}; \quad F_i^n(t) = 1 - e^{-\theta_i^n t};$$

$$L_1(t) = L_2(t) = 1 - e^{-\alpha t}, \quad \alpha \geq 0;$$

$$\gamma = c = 1; \quad \rho_i \text{ arbitrary},$$

for various values of α; α measures the degree of censoring present. These plots are time transformations of the more general case F continuous,

$$1 - F_i^n = (1-F)^{\theta_i^n}; \quad 1 - L_1 = 1 - L_2 = (1-F)^{\alpha}; \quad \gamma = c = 1.$$

Note that the test statistic of EFRON is "accidentally" optimal at $\alpha = 1$ when $k_C = k_E$, and that $e(k_E,t)$ is near zero for large t for $\alpha \geq 3$, when (4.3.18) fails. Again, the advisability of "stopping" the test statistic of EFRON earlier than the last observation is apparent.

The fact that

$$\left|\frac{dG_i^n}{dG} - 1\right| \leq \left|\theta_i^n - 1\right|$$

makes it very easy to verify, under H_1, Conditions II, II*, III, III*, (4.3.4) and (4.3.5) for the test statistics of GEHAN, EFRON and COX in suitable modifications of Propositions 4.3.1 to 4.3.3; we omit the details.

We now compare the test statistic of COX with the most powerful test for this problem. In the model specified by (5.2.19) to (5.2.21), let us suppose that for each n, the likelihood-ratio test statistic based on the observations $(\tilde{X}_{ij}^n, \delta_{ij}^n)$, $j = 1,\ldots,n_i$, $i = 1,2$ for testing H_0: $F_1^n = F_2^n = F$ (i.e. $c = 0$) versus H_1: "c is fixed and non-zero" is of the form given by Theorem 3.1.2:

(5.2.33)

$$\frac{dP_1}{dP_0} = \prod_{i,j:\delta_{ij}^n=1} \frac{1-\Delta G(\tilde{X}_{ij}^n)}{1-\Delta G_i^n(\tilde{X}_{ij}^n)} \frac{dG_i^n}{dG}(\tilde{X}_{ij}^n) \prod_{i,j} \frac{1-F_i^n(\tilde{X}_{ij}^n)}{1-F(\tilde{X}_{ij}^n)}$$

$$= \prod_{i,j:\delta_{ij}^n=1} \theta_i^n \cdot \prod_{i,j} \left\{\left(\prod_{s\leq\tilde{X}_{ij}^n} \frac{1-\Delta G_i^n(s)}{1-\Delta G(s)} \exp((\theta_i^n-1)\Delta G(s))\right)\exp(-(\theta_i^n-1)G(\tilde{X}_{ij}^n))\right\}.$$

Here we have used (3.2.9) and the fact that by (5.2.19)

Efficacies $e(k,t)$ with $k = k_C, k_G$ and k_E; $F(t) = 1 - \exp(-t)$; $\gamma = 1$ (Lehmann alternatives); and $1 - L_1 = 1 - L_2 = (1 - F)^{\alpha}$

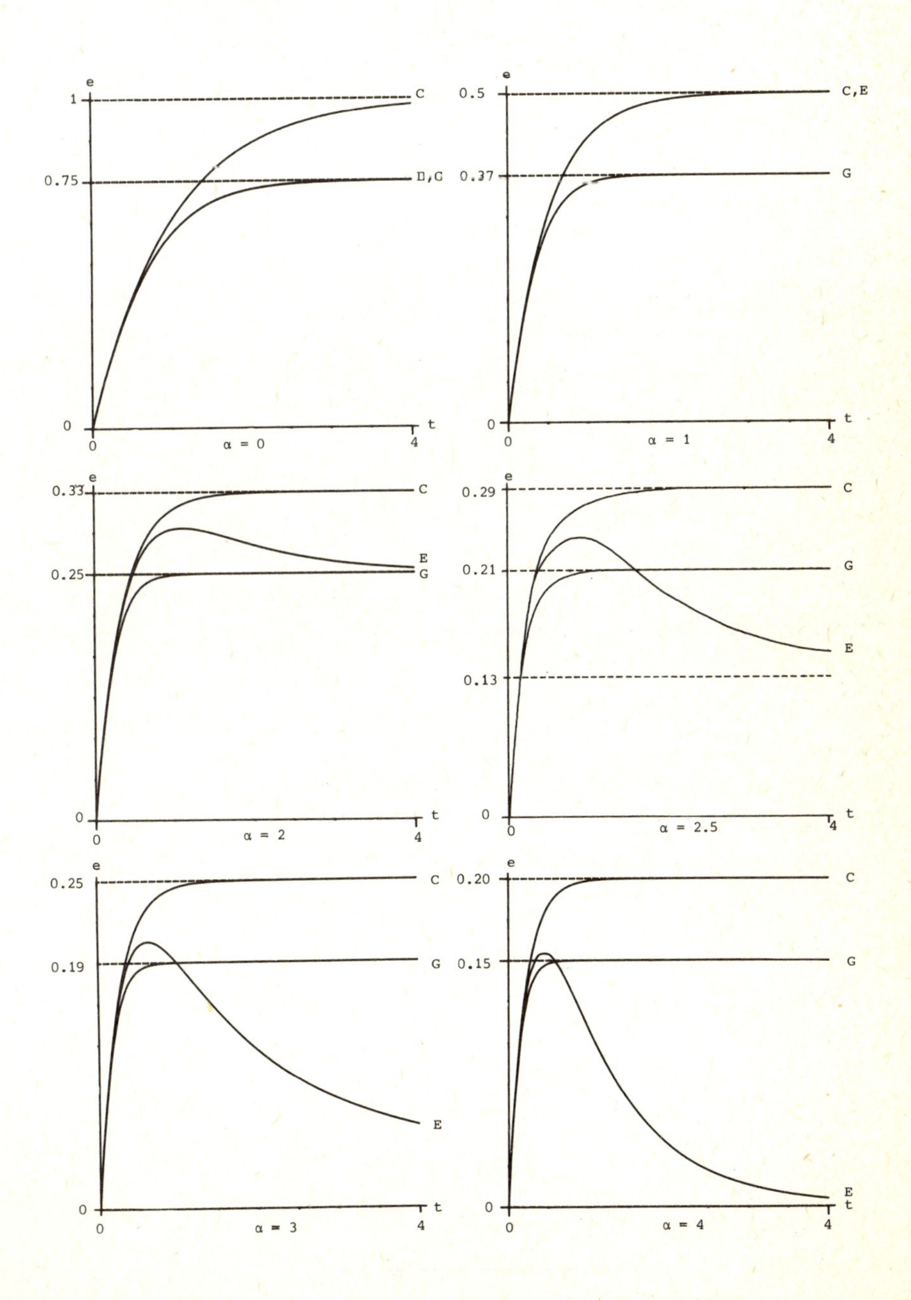

Figure 5.2.1.

$$-G^n_{ic}(t) + G_c(t) = \int_0^t (dG_c - dG^n_{ic}) = -\int_0^t (\theta^n_i - 1)\,dG_c$$

$$= -(\theta^n_i - 1)G(t) + \sum_{s\leq t} (\theta^n_i - 1)\Delta G(s).$$

Thus

$$(5.2.24) \qquad \log \frac{dP_1}{dP_0} = \sum_{i=1}^{2} \left(N_i(\infty) \log \theta^n_i - (\theta^n_i - 1) \int_0^\infty Y_i dG + B_i \right)$$

$$= \sum_{i=1}^{2} (\log(\theta^n_i) M_i(\infty) - A_i + B_i),$$

where $M_i = N_i - \int Y_i dG$ as we are working under H_0, where

$$(5.2.25) \qquad B_i = \sum_{j=1}^{n_i} \sum_{s\leq \tilde{X}^n_{ij}} \left\{ (\theta^n_i - 1)\Delta G(s) + \log\left(\frac{1 - \Delta G^n_i(s)}{1 - \Delta G(s)} \right) \right\}$$

and where

$$(5.2.26) \qquad A_i = ((\theta^n_i - 1) - \log \theta^n_i) \int_0^\infty Y_i dG.$$

We shall show that under H_0, and under (5.2.4) and (5.2.5), the following relationships hold (all limits being taken as $n \to \infty$):

$$(5.2.27) \qquad \log(\theta^n_i) M_i(\infty) \to_{\mathcal{D}} \mathcal{N}\left(0, c^2 \rho_{i'} \int_0^\infty y_i (1 - \Delta G)\,dG\right) \qquad (i' \neq i),$$

with $\log(\theta^n_1) M_1(\infty)$ and $\log(\theta^n_2) M_2(\infty)$ asymptotically independent,

$$(5.2.28) \qquad A_i \to_P \frac{c^2}{2} \rho_{i'} \int_0^\infty y_i dG,$$

and

$$(5.2.29) \qquad B_i \to_P \frac{c^2}{2} \rho_{i'} \int_0^\infty y_i \Delta G\,dG,$$

so that under (5.2.23)

$$(5.2.30) \qquad \log \frac{dP_1}{dP_0} \to_{\mathcal{D}} \mathcal{N}(-\tfrac{1}{2} c^2 \sigma^2_L, c^2 \sigma^2_L)$$

with

$$(5.2.31) \qquad \sigma^2_L = \int_0^\infty (\rho_1 y_2 + \rho_2 y_1)(1 - \Delta G)\,dG$$

(L standing for likelihood ratio):

<u>THEOREM 5.2.1</u>. *Suppose that* (5.2.4) *and* (5.2.5) *hold. If the likelihood ratio for the alternative hypothesis* H_1 *specified by* (5.2.19) *versus* H_0 *is given by* (5.2.23), *then under* H_0 (5.2.30) *holds with* σ_L^2 *defined by* (5.2.31).

<u>PROOF</u>. We first establish (5.2.27) and the asymptotic independence of $\log(\theta_1^n)M_1(\infty)$ and $\log(\theta_2^n)M_2(\infty)$. We shall continually use the expansion

$$\log(1+x) = x - \frac{x^2}{2} + O(x^3) \quad \text{as } x \to 0.$$

Thus we can write

$$\log(\theta_i^n)M_i(\infty) = \pm c(1+O(n_i^{-\frac{1}{2}}))\sqrt{\frac{n_{i'}}{n_1+n_2}}\int_0^\infty n_i^{-\frac{1}{2}}\,dM_i, \quad i' \neq i,$$

and we now apply a version of Theorem 4.2.1 with $H_i = n_i^{-\frac{1}{2}}$. Let us define $I_i = \{t\colon y_i(t) > 0\}$ and $u_i = \sup I_i$. As was remarked after the proof of Theorem 4.2.1, the theorem also holds with I depending on i if the conclusion is modified appropriately. With the interval I_i in place of I, with $H_i = n_i^{-\frac{1}{2}}$ and $h_i = y_i$, Conditions I and IIa follow immediately. Condition IIb also holds because if $u_i \notin I_i$, then by Lemma 4.3.2

$$\int_{(t,u_i]} H_i^2 Y_i\,dG = \int_{(t,u_i]} \frac{Y_i}{n_i}dG \to_P \int_{(t,u_i]} y_i\,dG \downarrow 0$$

as $t \uparrow u$, while similarly Condition III holds because

$$\int_{(u_i,\infty)} H_i^2 Y_i\,dG = \int_{(u_i,\infty)} \frac{Y_i}{n_i}dG \to_P 0 \quad \text{as } n \to \infty.$$

Next we consider A_i. By the expansion for log(1+x) given above, we have

$$A_i = \frac{c^2}{2}(1+O(n_i^{-\frac{1}{2}}))\,\frac{n_{i'}}{n_1+n_2}\int_0^\infty \frac{Y_i}{n_i}dG \to_P \frac{c^2}{2}\rho_{i'}\int_0^\infty y_i\,dG$$

by Lemma 4.3.2.

Finally we prove (5.2.29). By the arguments just after formula (5.2.21), successively substituting for $(1-\Delta G_i^n)/(1-\Delta G)$,

$$\frac{1-\Delta G_i^n}{1-\Delta G} = 1 - \frac{\Delta G_i^n - \Delta G}{1-\Delta G} = 1 - (\theta_i^n - 1)\Delta G\,\frac{1-\Delta G_i^n}{1-\Delta G}$$

$$= 1 - (\theta_i^n-1)\Delta G + (\theta_i^n-1)^2(\Delta G)^2\frac{1-\Delta G_i^n}{1-\Delta G} =$$

$$= 1 - (\theta_i^n - 1)\Delta G + (\theta_i^n - 1)^2 (\Delta G)^2 - (\theta_i^n - 1)^3 (\Delta G)^3 \frac{1 - \Delta G_i^n}{1 - \Delta G}$$

$$= 1 - (\theta_i^n - 1)\Delta G + (\theta_i^n - 1)^2 (\Delta G)^2 - (\theta_i^n - 1)^3 (\Delta G)^2 \frac{\Delta G_i^n}{\theta_i^n} .$$

Thus by the expansion of the logarithm, as $n \to \infty$,

$$(\theta_i^n - 1)\Delta G + \log\left(\frac{1 - \Delta G_i^n}{1 - \Delta G}\right) = \tfrac{1}{2}(\theta_i^n - 1)^2 \Delta G^2 + \Delta G^2 \, \mathcal{O}(|\theta_i^n - 1|^3)$$

and hence

$$B_i = \tfrac{1}{2}(\theta_i^n - 1)^2 \int_0^\infty Y_i \Delta G dG (1 + \mathcal{O}(|\theta_i^n - 1|))$$

$$= \frac{c^2}{2} \frac{n_{i'}}{n_1 + n_2} \int_0^\infty \frac{Y_i}{n_i} \Delta G dG (1 + \mathcal{O}(n_i^{-\frac{1}{2}})) \qquad i' \neq i$$

$$\to_P \frac{c^2}{2} \rho_{i'} \int_0^\infty Y_i \Delta G dG \qquad \text{as } n \to \infty$$

using Lemma 4.3.2 to extend convergence of $\int_0^t \frac{Y_i}{n_i} \Delta G dG$ for $t \in I_i$ to $t = \infty$. □

Now we have already shown that under H_0 and under the Conditions II and III for the test statistic of COX we have

(5.2.32) $\quad W_C(\infty) \to_{\mathcal{D}} N(0, \sigma_C^2)$

(5.2.33) $\quad V_{C\ell}(\infty) \to_P \sigma_C^2 \qquad \ell = 1$ or 2,

while under H_1 and the Conditions II, III, II* and III* we have

(5.2.34) $\quad W_C(\infty) \to_{\mathcal{D}} N(c\sigma_C^2, \sigma_C^2)$,

where

(5.2.35) $\quad \sigma_C^2 = \int_0^\infty \frac{Y_1 Y_2}{\rho_1 Y_1 + \rho_2 Y_2} (1 - \Delta G) dG.$

Now by Le Cam's first lemma (see e.g. HÁJEK & ŠIDÁK (1967)), (5.2.30) and (5.2.32) imply that (5.2.33) also holds under H_1, so we need not verify (4.3.4) and (4.3.5) under H_1 for the case $\ell = 2$. By Le Cam's third lemma, (5.2.30) implies that under H_1

$$\log \frac{dP_1}{dP_0} \to_{\mathcal{D}} N(\tfrac{1}{2}c^2\sigma_L^2, c^2\sigma_L^2).$$

So under H_1,

$$(c\sigma_L)^{-1}\left(\log \frac{dP_1}{dP_0} + \tfrac{1}{2}c^2\sigma_L^2\right) \to_{\mathcal{D}} N(c\sigma_L, 1)$$

and

$$\frac{W_C(\infty)}{\sqrt{V_{C\ell}(\infty)}} \to_{\mathcal{D}} N(c\sigma_C, 1);$$

under H_0 the same relationships hold with limiting means zero. Thus the efficacies of the test statistic of COX and the likelihood ratio test are $c^2\sigma_C^2$ and $c^2\sigma_L^2$ respectively, and hence the asymptotic relative efficiency of the former with respect to the latter is given by

$$\frac{\sigma_C^2}{\sigma_L^2} = \frac{\int_I (\rho_1 y_2^{-1} + \rho_2 y_1^{-1})^{-1} (1 - \Delta G)\,dG}{\int_0^\infty (\rho_1 y_2 + \rho_2 y_1)(1 - \Delta G)\,dG}.$$

Now on I

$$\begin{aligned}(\rho_1 y_2 + \rho_2 y_1)(\rho_1 y_2^{-1} + \rho_2 y_1^{-1}) &= \rho_1^2 + \rho_1\rho_2(y_1 y_2^{-1} + y_2 y_1^{-1}) + \rho_2^2 \\ &= 1 + \rho_1\rho_2(y_1 y_2^{-1} + y_2 y_1^{-1} - 2) \\ &= 1 + \rho_1\rho_2\left(\sqrt{\frac{y_1}{y_2}} - \sqrt{\frac{y_2}{y_1}}\right)^2 \geq 1.\end{aligned}$$

This gives us

COROLLARY 5.2.1. *The test statistic of COX is asymptotically most powerful against the alternatives* (5.2.19) *if and only if* $\rho_i = 0$ *and* $y_i = 0$ *outside* I *almost everywhere* $-dF$ *where* $\Delta G < 1$ *for* $i = 1$ *or* 2, *or if* $y_1 = y_2$ *almost everywhere* $-dF$ *where* $\Delta G < 1$.

This behaviour can be intuitively understood as follows. Under the simplest type of censoring, Example 4.1.1 with $L_{ij}^n = \chi_{[u_i,\infty)}$ for all i, j and n (Type I censorship in each sample apart), the result states that if $\rho_i \in (0,1)$ we have efficiency 1 iff $u_1 = u_2$. Both the likelihood ratio test (for which F must be known) and the test based on the test statistic of COX can be thought of as comparing estimates of F_1^n and F_2^n. If $u_1 < u_2$, the test statistic of COX only uses the information of what happens on $[0,u_1]$; because F being arbitrary, the available information about F_2^n based on what happens in $(u_1,u_2]$ is of no use. However the likelihood ratio test statistic, for which F must be known, can use the information of what happens in $(u_1,u_2]$ to improve its estimate of F_2^n (via an improved estimation of θ_2^n) and hence make a better comparison of F_1^n and F_2^n. What is remarkable is rather the fact that if $u_1 = u_2$, both tests are asymptotically equally good. We suggest that this behaviour is inherited by more complicated types of censoring; since the asymptotic results only depend on the limiting average censoring distributions, which might just as well have come about from the censoring of Example 4.1.1 with $L_{ij}^n = \chi_{[u_{ij}^n,\infty)}$ - a mixture of the type that has just been considered - this is hardly surprising. We see too that this behaviour should not depend on the special alternative hypothesis considered here. In a slightly different context AALEN (1976) sketches an application of results in LE CAM (1960) which shows that even if $y_1 \neq y_2$, the test statistic of COX is asymptotically uniformly most powerful against Lehmann alternatives in the class of asymptotically similar tests. Here F is considered as the nuisance parameter so that intuitively speaking the classes of similar tests and nonparametric tests coincide. The method of proof can be adapted to our situation, and also applies to the optimal tests of the class K discussed in the next section.

Finally we note that under (5.2.23), we could also have derived (5.2.34) by considering the joint asymptotic distribution of $\log \frac{dP_1}{dP_0}$ and $W_C(\infty)$, and then applying Le Cam's third lemma. Since both statistics can be written as stochastic integrals with respect to M_1 and M_2 (apart from the terms in $\log \frac{dP_1}{dP_0}$ which converge in probability to constants) this is a perfectly feasible approach; we could apply the Cramèr-Wold device and consider arbitrary linear combinations of $\int \frac{K}{Y_i} dM_i$ with $n_i^{-\frac{1}{2}} M_i$, i = 1 and 2, in order to be able to use Theorem 4.2.1. We shall use an argument along these lines in Section 5.3.

5.3. Optimal tests of the class K for parametric alternatives

We saw in the previous section that the optimal test statistics of the class K for testing against a contiguous sequence of alternatives for which (5.2.2) holds has limiting weight function

$$k \propto \frac{\gamma}{1-\Delta G} \frac{y_1 y_2}{\rho_1 y_1 + \rho_2 y_2}$$

and hence efficacy (when the test statistic is evaluated at time t)

$$e(k,t) = \int_0^t \frac{\gamma^2}{1-\Delta G} \frac{y_1 y_2}{\rho_1 y_1 + \rho_2 y_2} dG.$$

(We suppose throughout this section that (5.2.4) and (5.2.5) hold.) Now suppose that $\{F_\theta : \theta \in \Theta\}$ is some family of *continuous* distribution functions on $[0,\infty)$ indexed by a parameter θ taking values in a real interval Θ. We write as usual $G_\theta = \int (1-F_\theta)^{-1} dF_\theta$. Suppose the distribution functions under alternative and null hypothesis F_i^n and F of the last section are such that

$$(5.3.1) \qquad \begin{aligned} F_i^n &= F_{\theta_i^n} \qquad i = 1,2, \quad n = 1,2,\ldots \\ F &= F_{\theta_0} \end{aligned}$$

for some θ_0 and $\theta_i^n \in \Theta$. If F_θ has a density f_θ and hazard rate $\lambda_\theta = f_\theta(1-F_\theta)^{-1}$ with respect to some σ-finite measure μ, it is easy to see that

$$(5.3.2) \qquad \frac{dG_i^n}{dG}(t) = \frac{dG_{\theta_i^n}}{dG_{\theta_0}}(t) = \frac{\lambda_{\theta_i^n}(t)}{\lambda_{\theta_0}(t)} .$$

Therefore, defining γ_i by (5.2.2) if the limit there exists (even if convergence is not uniform), if for some fixed $c \neq 0$

$$(5.3.3) \qquad \theta_i^n = \theta_0 \pm c\sqrt{\frac{n_{i'}}{n_i(n_1+n_2)}} \qquad i' \neq i, \quad \pm = (-1)^{i+1},$$

and if $\lambda_\theta(t)$ is differentiable with respect to θ at $\theta = \theta_0$ for μ-almost all t, then

$$\gamma_i(t) = \pm c\rho_{i'} \frac{\partial}{\partial\theta} \log \lambda_\theta(t) \bigg|_{\theta=\theta_0}$$

and (cf. (5.2.3))

$$(5.3.4) \qquad \gamma(t) = \gamma_1(t) - \gamma_2(t) = c \frac{\partial}{\partial\theta} \log \lambda_\theta(t) \bigg|_{\theta=\theta_0}$$

for μ-almost t.

This suggests we should try to find test statistics in $\mathcal{K}$ for which $\sqrt{\frac{n_1+n_2}{n_1 n_2}}\, K$ converges under H_0 to

$$(5.3.5) \qquad k \propto \frac{\partial}{\partial\theta} \log \lambda_\theta \bigg|_{\theta=\theta_0} \frac{y_1 y_2}{\rho_1 y_1 + \rho_2 y_2}$$

whatever the value of θ_0 or the limiting average censoring distributions L_1 and L_2; such a test statistic should have efficacy

$$(5.3.6) \qquad e(k,t) = c^2 \int_0^\infty \left(\frac{\partial}{\partial\theta} \log \lambda_\theta \bigg|_{\theta=\theta_0}\right)^2 \frac{y_1 y_2}{\rho_1 y_1 + \rho_2 y_2}\, dG$$

and be optimal in K for the family $\{F_\theta : \theta \in \Theta\}$.

The following proposition shows once more that such a test statistic will only have a Pitman asymptotic relative efficiency of 100% with respect to the most powerful test against the alternatives specified by (5.3.1) and (5.3.3) when either $y_1 = y_2$, or for i = 1 or 2, $\rho_i = 0$ and $y_i = 0$ where $y_{i'} = 0$ $(i' \neq i)$:

PROPOSITION 5.3.1. *Suppose that* F_i^n *and* F *are given by* (5.3.1) *and* (5.3.3), *that* $\log dP_1/dP_0$ *is given by* (3.1.8) *for each* n, *and that* (5.2.4) *and* (5.2.5) *hold under* H_0. *Suppose also that* $\frac{\partial}{\partial\theta} \log \lambda_\theta(t)\big|_{\theta=\theta_0}$ *exists almost everywhere* $-dF_{\theta_0}(t)$ *and that*

$$(5.3.7) \qquad \lim_{\theta\to\theta_0} \int_0^\infty \left(\frac{2}{\lambda_{\theta_0}^{\frac{1}{2}}} \frac{\lambda_\theta^{\frac{1}{2}} - \lambda_{\theta_0}^{\frac{1}{2}}}{\theta - \theta_0}\right)^2 dF_{\theta_0} = \int_0^\infty \left(\frac{\partial}{\partial\theta} \log \lambda_\theta \bigg|_{\theta=\theta_0}\right)^2 dF_{\theta_0} < \infty.$$

Then under H_0

$$\log \frac{dP_1}{dP_0} \to_{\mathcal{D}} N(-\tfrac{1}{2}c^2\sigma_L^2, c^2\sigma_L^2)$$

where

$$\sigma_L^2 = \int_0^\infty (\rho_1 y_2 + \rho_2 y_1)\left(\frac{\partial}{\partial\theta} \log \lambda_\theta \bigg|_{\theta=\theta_0}\right)^2 dG_{\theta_0} < \infty.$$

PROOF. Since F_θ is continuous for all θ, by (3.1.8) and (3.2.9) we can write

$$(5.3.8) \qquad \log \frac{dP_1}{dP_0} = \sum_i \left(\sum_{j:\delta_{ij}=1} \log \frac{dG_{\theta_i^n}}{dG_{\theta_0}}(\tilde{X}_{ij}^n) - \sum_j (G_{\theta_i^n}(\tilde{X}_{ij}^n) - G_{\theta_0}(\tilde{X}_{ij}^n))\right)$$

$$= \sum_i \left(\int_0^\infty \log \frac{dG_{\theta_i^n}}{dG_{\theta_0}} dN_i - \int_0^\infty \left(\frac{dG_{\theta_i^n}}{dG_{\theta_0}} - 1\right) Y_i dG_{\theta_0}\right)$$

$$= \sum_i \int_0^\infty 2\left(\sqrt{\frac{\lambda_{\theta_i^n}}{\lambda_{\theta_0}}} - 1\right) dM_i - \tfrac{1}{2} \sum_i \int_0^\infty \left(2\left(\sqrt{\frac{\lambda_{\theta_i^n}}{\lambda_{\theta_0}}} - 1\right)\right)^2 Y_i dG +$$

$$+ 2\sum_i \int_0^\infty \left(\log \sqrt{\frac{\lambda_{\theta_i^n}}{\lambda_{\theta_0}}} - \left(\sqrt{\frac{\lambda_{\theta_i^n}}{\lambda_{\theta_0}}} - 1\right) - \tfrac{1}{2}\left(\sqrt{\frac{\lambda_{\theta_i^n}}{\lambda_{\theta_0}}} - 1\right)^2\right) dN_i$$

$$+ \sum_i \int_0^\infty \left(\sqrt{\frac{\lambda_{\theta_i^n}}{\lambda_{\theta_0}}} - 1\right)^2 Y_i dG - \sum_i \int_0^\infty \left(\sqrt{\frac{\lambda_{\theta_i^n}}{\lambda_{\theta_0}}} - 1\right)^2 dN_i.$$

Here M_i is defined by $M_i = N_i - \int Y_i dG$ (and not dG_i^n) as we are working under H_0. Let us define

$$Z_i^n = \int 2\left(\sqrt{\frac{\lambda_{\theta_i^n}}{\lambda_{\theta_0}}} - 1\right) dM_i$$

and

$$\bar{Z}_i^{n\varepsilon} = \int 2\left(\sqrt{\frac{\lambda_{\theta_i^n}}{\lambda_{\theta_0}}} - 1\right) \chi_{\left\{2\left|\sqrt{\frac{\lambda_{\theta_i^n}}{\lambda_{\theta_0}}} - 1\right| > \varepsilon\right\}} dM_i.$$

Note that almost surely,

$$\sum_{s\in[0,\infty)} (\tfrac{1}{2}\Delta Z_i^n(s))^2 = \int_0^\infty \left(\sqrt{\frac{\lambda_{\theta_i^n}}{\lambda_{\theta_0}}} - 1\right)^2 dN_i.$$

With continuous F, M_1 and M_2 never jump simultaneously and $\sup_{[0,\infty)} |\Delta M_i| \le 1$. So $\{\bar{Z}_i^{n\varepsilon}: i = 1,2\}$ forms the jump part of an ε-decomposition of $\{Z_i^n: i = 1,2\}$, and by Theorem 2.4.1 (making use of (2.4.9) to deal with the last two terms of (5.3.8)) it suffices to show that

$$(5.3.9) \qquad \langle Z_i^n, Z_i^n\rangle(t) =$$

$$= \int_0^t \left(2\left(\sqrt{\frac{\lambda_{\theta_i^n}}{\lambda_{\theta_0}}} - 1\right)\right)^2 Y_i dG \to_P c^2 \int_0^t \rho_{i'}Y_i \left(\frac{\partial}{\partial\theta}\log \lambda_\theta\Big|_{\theta=\theta_0}\right)^2 dG < \infty$$

for all $t \in [0,\infty]$, that

$$(5.3.10) \qquad \langle \bar{Z}_i^{n\varepsilon}, \bar{Z}_i^{n\varepsilon}\rangle(\infty) = \int_0^\infty \left(2\left(\sqrt{\frac{\lambda_{\theta_i^n}}{\lambda_{\theta_0}}} - 1\right)\right)^2 \chi_{\left\{2\left|\sqrt{\frac{\lambda_{\theta_i^n}}{\lambda_{\theta_0}}} - 1\right| > \varepsilon\right\}} Y_i dG \to_P 0$$

for all $\varepsilon > 0$, and that

$$(5.3.11) \qquad \int_0^\infty \left(\log\sqrt{\frac{\lambda_{\theta_i^n}}{\lambda_{\theta_0}}} - \left(\sqrt{\frac{\lambda_{\theta_i^n}}{\lambda_{\theta_0}}} - 1\right) + \tfrac{1}{2}\left(\sqrt{\frac{\lambda_{\theta_i^n}}{\lambda_{\theta_0}}} - 1\right)^2\right) dN_i \to_P 0,$$

all as $n \to \infty$. Now finiteness of the right hand side of (5.3.9) follows from the finiteness assertion in (5.3.7) since $y_i dG \le dF$. By the equality

$$\log x = (x-1) - \tfrac{1}{2}(x-1)^2 \int_0^1 2(1-z)(1+z(x-1))^{-2} dz$$

(this equality is used in the proof of Le Cam's second lemma, see e.g. HÁJEK & ŠIDÁK (1967) page 206), (5.3.11) is equivalent to

$$(5.3.12) \qquad \int_0^\infty \left(\sqrt{\frac{\lambda_{\theta_i^n}}{\lambda_{\theta_0}}} - 1\right)^2 \left[1 - \int_0^1 (1-z)\left(1+z\left(\sqrt{\frac{\lambda_{\theta_i^n}}{\lambda_{\theta_0}}} - 1\right)\right)^{-2} dz\right] dN_i \to_P 0 \quad \text{as } n \to \infty.$$

Let us assume that (5.3.9) and (5.2.10) hold, so that by Theorem 2.4.1 the martingales $Z_i^n = \int 2(\sqrt{\lambda_{\theta_i^n}/\lambda_{\theta_0}} - 1)dM_i$ converge weakly in $D[0,\infty]$ to a continuous limit as $n \to \infty$. It then follows that the suprema over $[0,\infty]$ of the absolute value of the jumps of these martingales converge in probability to zero; i.e.

$$\sup_{[0,\infty]} \left|\sqrt{\frac{\lambda_{\theta_i^n}}{\lambda_{\theta_0}}} - 1\right| \Delta N_i \to_P 0 \quad \text{as } n \to \infty, \quad i = 1 \text{ and } 2.$$

On the event where this supremum is less than ε, the left hand side of (5.3.12) is smaller in absolute value than

$$2\varepsilon \int_0^\infty \left(\sqrt{\frac{\lambda_{\theta_i^n}}{\lambda_{\theta_0}}} - 1\right)^2 dN_i.$$

So under (5.3.9) and (5.3.10), (5.3.11) holds if

$$(*) \qquad \int_0^\infty \left(\sqrt{\frac{\lambda_{\theta_i^n}}{\lambda_{\theta_0}}} - 1\right)^2 dN_i$$

is bounded in probability as $n \to \infty$. But this also follows from (5.3.9) and (5.3.10), because then as we remarked earlier by (2.4.9), (*) converges in probability to the (finite) limit in probability of

$$\int_0^\infty \left(\sqrt{\frac{\lambda_{\theta_i^n}}{\lambda_{\theta_0}}} - 1\right)^2 Y_i dG.$$

It suffices therefore to verify (5.3.9) and (5.3.10). Now by the well-known Hájek lemma (HÁJEK & ŠIDÁK (1967) page 154), (5.3.7) implies that

$$\int_0^\infty \left(\frac{2}{\lambda_{\theta_0}^{1/2}} \frac{\lambda_\theta^{1/2} - \lambda_{\theta_0}^{1/2}}{\theta - \theta_0} - \frac{\partial}{\partial\theta} \log \lambda_\theta \Big|_{\theta=\theta_0} \right)^2 dF \to 0 \quad \text{as } \theta \to \theta_0.$$

We can rewrite the left hand side of (5.3.9) as

$$c^2 \frac{n_{i'}}{n_1+n_2} \int_0^t \left(\frac{2}{\lambda_{\theta_0}^{1/2}} \frac{\lambda_{\theta_i^n}^{1/2} - \lambda_{\theta_0}^{1/2}}{\theta_i^n - \theta_0} \right)^2 \frac{Y_i}{n_i}\, dG.$$

By VAN ZUIJLEN (1978) Theorem 1.1 and Corollary 3.1, for given $\varepsilon > 0$ there exists $\beta \in (0,1)$ such that under H_0,

$$P\left(\frac{Y_i}{n_i} \le \beta^{-1}(1-F) \text{ on } [0,\infty) \right) \ge 1-\varepsilon$$

uniformly in n_i. Let $\delta > 0$ and $s \in (0,\infty)$ be fixed. On the event where $\frac{Y_i}{n_i} \le \beta^{-1}(1-F)$ and $\sup\left|\frac{Y_i}{n_i} - y_i\right| \le \delta$ we have, for any $t \in (0,\infty]$,

$$\left| \int_0^t \left(\frac{2}{\lambda_{\theta_0}^{1/2}} \frac{\lambda_{\theta_i^n}^{1/2} - \lambda_{\theta_0}^{1/2}}{\theta_i^n - \theta_0} \right)^2 \frac{Y_i}{n_i}\, dG - \int_0^t \left(\frac{\partial}{\partial\theta} \log \lambda_\theta \Big|_{\theta=\theta_0} \right)^2 y_i\, dG \right|$$

$$\le \left| \int_0^{s\wedge t} \left(\frac{2}{\lambda_{\theta_0}^{1/2}} \frac{\lambda_{\theta_i^n}^{1/2} - \lambda_{\theta_0}^{1/2}}{\theta_i^n - \theta_0} \right)^2 y_i\, dG - \int_0^{s\wedge t} \left(\frac{\partial}{\partial\theta} \log \lambda_\theta \Big|_{\theta=\theta_0} \right)^2 y_i\, dG \right|$$

$$+ \frac{\delta}{1-F(s)} \int_0^{s\wedge t} \left(\frac{2}{\lambda_{\theta_0}^{1/2}} \frac{\lambda_{\theta_i^n}^{1/2} - \lambda_{\theta_0}^{1/2}}{\theta_i^n - \theta_0} \right)^2 dF$$

$$+ \beta^{-1} \int_{s\wedge t}^t \left(\frac{2}{\lambda_{\theta_0}^{1/2}} \frac{\lambda_{\theta_i^n}^{1/2} - \lambda_{\theta_0}}{\theta_i^n - \theta_0} \right)^2 dF + \int_{s\wedge t}^t \left(\frac{\partial}{\partial\theta} \log \lambda_\theta \Big|_{\theta=\theta_0} \right)^2 dF.$$

If s is chosen large enough subject to $F(s) < 1$, the last term is arbitrarily small and the last but one term converges to an arbitrarily small quantity as $n \to \infty$. The first term converges to zero as $n \to \infty$ (convergence in L^2-norm implies convergence of L^2-norms). The remaining term, involving δ, converges as $n \to \infty$ to an arbitrarily small quantity if δ is chosen small enough. Since ε was arbitrary and $P(\sup\left|\frac{Y_i}{n_i} - y_i\right| \le \delta) \to 1$ as $n \to \infty$, (5.3.9) holds. The relation (5.3.10) can be established in exactly the same way since

$$\int_0^\infty \left(\frac{2}{\lambda_{\theta_0}^{\frac{1}{2}}} \frac{\lambda_{\theta_i^n}^{\frac{1}{2}} - \lambda_{\theta_0}^{\frac{1}{2}}}{\theta_i^n - \theta_0} \right)^2 \chi_{\left\{ 2 \left| \sqrt{\frac{\lambda_{\theta_i^n}}{\lambda_{\theta_0}}} - 1 \right| > \varepsilon \right\}} dF$$

$$\leq 2 \int_0^\infty \left(\frac{2}{\lambda_{\theta_0}^{\frac{1}{2}}} \frac{\lambda_{\theta_i^n}^{\frac{1}{2}} - \lambda_{\theta_0}^{\frac{1}{2}}}{\theta_i^n - \theta_0} - \frac{\partial}{\partial\theta} \log \lambda_\theta \Big|_{\theta=\theta_0} \right)^2 \chi_{\left\{ 2 \left| \sqrt{\frac{\lambda_{\theta_i^n}}{\lambda_{\theta_0}}} - 1 \right| > \varepsilon \right\}} dF$$

$$+ 2 \int_0^\infty \left(\frac{\partial}{\partial\theta} \log \lambda_\theta \Big|_{\theta=\theta_0} \right)^2 \chi_{\left\{ 2 \left| \sqrt{\frac{\lambda_{\theta_i^n}}{\lambda_{\theta_0}}} - 1 \right| > \varepsilon \right\}} dF$$

$\to 0$ as $n \to \infty$. □

The above proof is very similar to the usual proof of Le Cam's second lemma. For instance, the proof of asymptotic negligeability of the remainder terms in (5.3.8) (i.e. proving that (5.3.11) holds) uses a consequence of asymptotic normality of the leading term; the same argument is used in Le Cam's second lemma too.

By Le Cam's third lemma, under the conditions of Proposition 5.3.1 we have under H_1

$$\log \frac{dP_1}{dP_0} \to_{\mathcal{D}} N(\tfrac{1}{2}c^2\sigma_L^2, c^2\sigma_L^2)$$

and hence the efficiency of the optimal test in K (whose efficacy is given by (5.3.6)) relative to the most powerful test against H_1 is

$$\frac{\int_0^\infty \left(\frac{\partial}{\partial\theta} \log \lambda_\theta \Big|_{\theta=\theta_0} \right)^2 \frac{y_1 y_2}{\rho_1 y_1 + \rho_2 y_2} dG}{\int_0^\infty \left(\frac{\partial}{\partial\theta} \log \lambda_\theta \Big|_{\theta=\theta_0} \right)^2 (\rho_1 y_2 + \rho_2 y_1) dG} \leq 1$$

with equality when $y_1 = y_2$, or for $i = 1$ or 2 $\rho_i = 0$ and $y_i = 0$ where $y_{i'} = 0$.

However it still remains to show that a test statistic in K can be constructed for which (5.3.5) holds and hence (5.3.6) does too. We shall only do this in the special situation in which

(5.3.13) $F_\theta(t) = \Psi(g(t)+\theta) \quad t \in [0,\infty), \quad \theta \in \Theta = (-\infty,\infty),$

where g is a fixed continuous nondecreasing function from $[0,\infty]$ onto $[-\infty,\infty]$ and Ψ is a fixed continuous distribution function with positive density ψ on $(-\infty,\infty)$, such that ψ', the derivative of ψ, exists and is continuous at all but finitely many points. We define $\lambda = \psi/(1-\Psi)$ and $\ell = \log \lambda$, and note that

(5.3.14) $\ell' = (\psi'/\psi) + \lambda$

exists where ψ' does. We suppose that except possibly on arbitrarily small neighbourhoods of at most finitely many points of $[-\infty,\infty]$, ℓ' is of bounded variation on $[-\infty,\infty]$. Finally we assume that according to some convention, ℓ' is assigned finite values in the points $\pm\infty$ and the points where ψ' does not exist.

The family defined by (5.3.13) might be termed a "time transformed location family". In fact θ is minus the location parameter for Ψ; the reason for this choice will become apparent shortly.

Now F_θ is continuous and has density $\psi(g(\cdot)+\theta)$ with respect to the σ-finite measure generated by g. Hence it has hazard rate $\lambda_\theta = \lambda(g+\theta)$ with respect to this measure. Since

$$\frac{\partial}{\partial\theta} \log \lambda_\theta(t) = \ell'(g(t)+\theta) = \ell'(\Psi^{-1}(F_\theta(t))),$$

in the hope that (5.3.5) holds, we define a test statistic in K by

(5.3.15) $$K = K_{opt} = \sqrt{\frac{n_1 n_2}{n_1+n_2}}\, \ell'(\Psi^{-1}(\hat{F}_-))\frac{Y_1}{n_1}\frac{Y_2}{n_2}\frac{n_1+n_2}{Y_1+Y_2},$$

where $\hat{F}$ is the product limit estimator of F_θ based on the combined sample. Possible alternatives could be to replace $\hat{F}$ in (5.3.15) with $\tilde{F} = (n\hat{F}+1)/(n+1)$, with $(n_1\hat{F}_1+n_2\hat{F}_2)/n$, or with $(n_1\hat{F}_1+n_2\hat{F}_2+1)/(n+1)$. The justification for (5.3.15) is that if $g(t)+\theta_0$ is not one of the points of discontinuity of ℓ', and if $y_1(t) > 0$ and $y_2(t) > 0$, then under H_0

$$\sqrt{\frac{n_1+n_2}{n_1 n_2}}\, K_{opt}(t) \to_P \ell'(\Psi^{-1}(F_{\theta_0}(t)))\, \frac{y_1(t)y_2(t)}{\rho_1 y_1(t)+\rho_2 y_2(t)}$$
$$= \ell'(g(t)+\theta_0)\, \frac{y_1(t)y_2(t)}{\rho_1 y_1(t)+\rho_2 y_2(t)}$$
$$= \frac{\partial}{\partial\theta} \log \lambda_\theta(t)\, \frac{y_1(t)y_2(t)}{\rho_1 t_1(t)+\rho_2 y_2(t)}\,.$$

In fact we have, in probability, uniform convergence on each compact interval on which $\ell'(g+\theta_0)$ is continuous and y_1 and y_2 are positive. The same holds for any of the alternatives to (5.3.15) mentioned above.

Let us note some other consequences of this definition. Firstly, K_{opt} is predictable, because Y_1, Y_2 and $\hat{F}_-$ are. Secondly, it is bounded, because for each n, $\hat{F}_-$ takes on values from some finite set of values and hence K_{opt} does too. Thirdly, neither θ_0 nor g enters into the specification of K_{opt}, as we required. Note that we need to define ℓ' in the point $-\infty$ because $\Psi^{-1}(\hat{F}_-) = -\infty$ at the first uncensored observation. K_{opt} is not necessarily nonnegative. However in cases in which shifting Ψ to the right decreases the hazard rate everywhere (such a shift can never increase it everywhere), ℓ' is nonnegative. This is why we chose to have $+\theta$ instead of $-\theta$ in (5.3.15). The following examples all have ℓ' nonnegative and nonincreasing, which means that the resulting test statistics are members of K^+ and hence should be consistent against alternatives of stochastic ordering (see Lemmas 4.1.6 and 4.1.7).

EXAMPLE 5.3.1. *Extreme value distribution (smallest extremes) of Type* I.

$$\Psi(x) = 1 - e^{-e^x}.$$

We find $\lambda(x) = e^x$ and $\ell'(x) = 1$, so that K_{opt} becomes simply K_C, the weight function for the test statistic of COX. This relationship is a reflection of the optimality of the test statistic of COX against Lehmann-alternatives,

$$(1 - F_i^n) = (1 - F)^{\alpha_i^n}$$

when F is continuous. For in this situation

$$F_i^n = 1 - \exp(\alpha_i^n \log(1 - F))$$

$$= \Psi(\log(-\log(1 - F)) + \log \alpha_i^n)$$

so that by taking $g = \log(-\log(1-F))$ and $\theta_i^n = \log \alpha_i^n$ we arrive at (5.3.13). Lehmann-alternatives arise for instance if F_θ is the exponential or Weibull distribution with scale parameter $\log(1/\theta)$.

EXAMPLE 5.3.2. *Logistic distribution.*

$$\Psi(x) = \frac{1}{1+e^{-x}}$$

$$\lambda(x) = \frac{1}{1+e^{-x}}$$

and

$$\ell'(x) = \frac{e^{-x}}{1+e^{-x}} = 1 - \Psi(x).$$

Making the natural definition $\ell'(-\infty) = 1$, we obtain

$$K_{opt} = (1-\hat{F}_-)K_C.$$

When there is no censoring, we find that

$$K_{opt} = K_G = K_E$$

and the three tests coincide with the Wilcoxon test based on the statistic $\int_0^\infty Y_2 dN_1 - \int_0^\infty Y_1 dN_2$. This is not unexpected: the test statistics of GEHAN and EFRON were constructed to be generalizations of the Wilcoxon test, which is asymptotically most powerful against contiguous location alternatives with the logistic distribution. In Figure 5.3.1 we plot e(k,t) for these alternatives in the same way as in Figure 5.2.1, including the new optimal test statistic.

EXAMPLE 5.3.3. *Double exponential distribution (Laplace distribution).*

$$\Psi(x) = \begin{cases} \tfrac{1}{2}e^{x} & x \le 0, \\ 1 - \tfrac{1}{2}e^{-x} & x \ge 0. \end{cases}$$

We find

$$\ell'(x) = \begin{cases} (1-\tfrac{1}{2}e^{x})^{-1} & x < 0, \\ 0 & x > 0, \end{cases}$$

so that defining $\ell'(-\infty) = 1$ and $\ell'(0) = 2$ we obtain

$$K_{opt} = (1-\hat{F}_-)^{-1} \cdot \chi_{[0,\frac{1}{2}]}(\hat{F}_-) \cdot K_C.$$

The resulting test statistic bears little resemblance to the sign test with

Efficacies $e(k,t)$ with $k = k_C, k_G, k_E$ and k_{opt}; $F(t) = 1 - \exp(-t)$;
$\gamma = 1 - F$ (logistic location alternatives); and $1 - L_1 = 1 - L_2 = (1 - F)^{\alpha}$

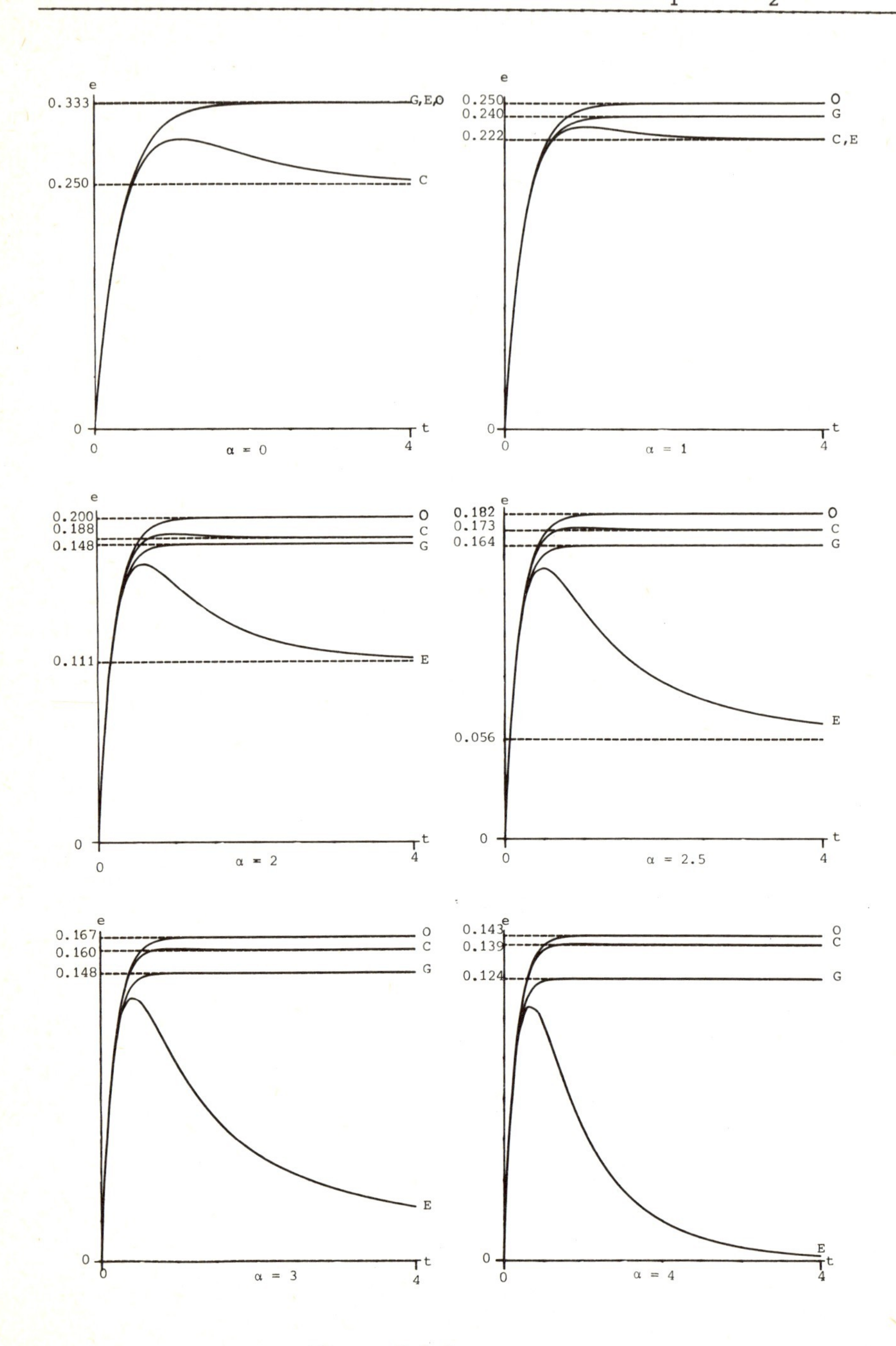

Figure 5.3.1.

which it should share asymptotic optimality properties when there is no censoring.

A similar optimal K is obtained if we take F_θ to be the uniform distribution on $[0,e^{-\theta}]$ so that

$$F_\theta(t) = e^{(\log t)+\theta}, \quad t \in [0,e^{-\theta}],$$

and we can set $g(t) = \log t$, $\Psi(x) = e^x$ on $(-\infty,0]$. This example conflicts with our requirement that ψ should be positive on $(-\infty,\infty)$; however if censoring is such that with probability 1 all observations are less that some fixed time $u < e^{-\theta_i^n}$ for all i and n, the test statistic defined by

$$K_{opt} = (1-\hat{F}_-)^{-1}K_C$$

will have the expected optimality properties.

EXAMPLE 5.3.4. *Normal distribution.*

$$\Psi(x) = \Phi(x)$$

where Φ is the standard normal distribution function with density ϕ. This covers the case in which F_θ is the lognormal distribution with parameters μ and σ such $\frac{\mu}{\sigma} = -\theta$ is the parameter of interest and σ, unknown, is the same in both samples (and hence can be absorbed into the transformation g). In this example, by (5.3.14),

$$\ell'(x) = -x+\lambda(x),$$

where

$$\lambda(x) = \phi(x)/(1-\Phi(x)).$$

It is well known that $\lambda(x)-x$ is positive for all x and $\lambda(x)-x \to 0$ as $x \to \infty$; obviously $\lambda(x) \to 0$ as $x \to -\infty$. So ℓ' is positive and $\ell'(x) \to \infty$ as $x \to -\infty$. Rather than assign ℓ' some arbitrary finite value in the point $-\infty$, it seems better to replace $\hat{F}$ in (5.3.15) with $\tilde{F} = (n\hat{F}+1)/(n+1)$, obtaining

$$K_{opt} = \left(-\Phi^{-1}(\tilde{F}_-) + \frac{\phi(\Phi^{-1}(\tilde{F}_-))}{1-\tilde{F}_-}\right)\cdot K_C.$$

The resulting test statistic has a completely different form from the test statistics of Van der Waerden or Fisher-Yates with respect to which it is asymptotically efficient when there is no censoring. There is no obvious way in which the latter statistics can be generalized to the case of censored data.

In a time transformed location family, Condition (5.3.7) of Proposition 5.3.1 is equivalent to

$$(5.3.16)\qquad \lim_{\delta\to 0}\int_{-\infty}^{\infty}\left(\frac{2}{\lambda^{\frac12}(x)}\,\frac{\lambda^{\frac12}(x+\delta)-\lambda^{\frac12}(x)}{\delta}\right)^2 d\Psi(x) = \int_{-\infty}^{\infty}(\ell')^2 d\Psi < \infty,$$

which can easily be verified for all the above examples. Note that

$$\begin{aligned}\int_{-\infty}^{t}(\ell')^2 d\Psi &= \int_{-\infty}^{t}\left(\frac{\psi'}{\psi}+\frac{\psi}{1-\Psi}\right)^2 d\Psi\\ &= \int_{-\infty}^{t}\left(\frac{\psi'}{\psi}\right)^2 d\Psi + 2\int_{-\infty}^{t}\frac{\psi'}{1-\Psi}\,d\Psi + \int_{-\infty}^{t}\left(\frac{\psi}{1-\Psi}\right)^2 d\Psi\\ &= \int_{-\infty}^{t}\left(\frac{\psi'}{\psi}\right)^2 d\Psi + \int_{-\infty}^{t}\left(\frac{\psi^2}{1-\Psi}\right)' dx\\ &= \int_{-\infty}^{t}\left(\frac{\psi'}{\psi}\right)^2 d\Psi + \frac{\psi(t)^2}{1-\Psi(t)},\end{aligned}$$

so that if $\lim_{t\to\infty}\psi(t)^2(1-\Psi(t))^{-1} = 0$, the limiting quantity in (5.3.16) equals the Fisher information for the location family $\{\Psi(\cdot+\theta):\ \theta\in(-\infty,\infty)\}$.

In proving asymptotic normality under the null hypothesis of the test statistic based on K_{opt}, the only essentially new difficulties occur when, as in the case $\Psi = \Phi$, an $x \in [-\infty,\infty]$ exists such that $\limsup_{y\to\infty}|\ell'(y)| = \infty$. (In this case, the function k defined in (5.2.7) does not have the usually required properties.)

In the following proposition, we suppose that $x = -\infty$ is the only such point (if any exists at all); however the conditions can be modified in a straightforward fashion to cover other cases. After this proposition, we give a result (Proposition 5.3.3) on the joint asymptotic normality of $\log\frac{dP_1}{dP_0}$ and $W_{opt}(\infty)$, from which the expected efficiency result is derived (corollary 5.3.1). Then we continue the discussion of Examples 5.3.1 to 5.3.4.

PROPOSITION 5.3.2 *(Asymptotic normality of* $W_{opt}(\infty)/\sqrt{V_{\ell opt}(\infty)}$ *under* H_0*). Let* Ψ*,* ℓ *and* ℓ' *have the properties given after* (5.3.13) *and define* K_{opt} *by* (5.3.15) *or by one of the alternatives given immediately afterwards.*

Suppose that $F_i^n = F$ *for all* i *and* n *for some continuous distribution function* F *and that* (5.2.4) *and* (5.2.5) *hold, and define*
$u = \sup\{t: y_1(t) \wedge y_2(t) > 0\}$. *If for* $i = 1$ *or* 2 $\rho_i = 0$, *suppose either that* $y_i(u+) = 0$ *or that for each* n, $Y_1(u+) \wedge Y_2(u+) = 0$ *almost surely. Suppose either that* ℓ' *has a limit in* $-\infty$ *and is bounded on* $(-\infty,\infty)$, *or alternatively that* ℓ' *is bounded on* $[x,\infty)$ *for each* $x > -\infty$,

$$(5.3.17) \qquad \int_{-\infty}^{\infty} \ell'^2 d\Psi < \infty,$$

and

$$(5.3.18) \qquad \lim_{t\downarrow 0} \limsup_{n\to\infty} P\left(\int_0^t \ell'(\Psi^{-1}(\hat{F}_-))^2 dF > \varepsilon\right) = 0$$

for all $\varepsilon > 0$ *(with* $\hat{F}$ *replaced by one of the alternatives as appropriate). Then the statistics defined in* (4.1.18), (4.1.20) *and* (4.1.21) *with* $K = K_{opt}$ *satisfy*

$$(5.3.19) \qquad W_{opt}(\infty) \to_{\mathcal{D}} N\left(0, \int_0^\infty \frac{y_1 y_2}{\rho_1 y_1 + \rho_2 y_2} (\ell'(\Psi^{-1}(F)))^2 dG\right)$$

and

$$(5.3.20) \qquad V_{\ell opt}(\infty) \to_P \int_0^\infty \frac{y_1 y_2}{\rho_1 y_1 + \rho_2 y_2} (\ell'(\Psi^{-1}(F)))^2 dG, \qquad \ell = 1 \text{ or } 2,$$

as $n \to \infty$.

PROOF. For each $r \in \mathbb{N}$ let $B_r \subset (0,\infty]$ be a finite union of intervals of the form (a,b] such that $\ell'(\Psi^{-1}(F))$ is continuous and of bounded variation outside B_r and such that $\{B_r: r = 1,2,\ldots\}$ forms a decreasing sequence of sets whose intersection is finite. In particular, B_r contains a subinterval $(0,t_r]$ where $t_r \downarrow 0$ as $r \to \infty$ if ℓ' does not have a limit in $-\infty$, or is not bounded on $(-\infty,\infty)$. Let B_r^c be the complement of B_r on $(0,\infty)$. It is easy to check that Condition I of Theorem 4.2.1 and Lemma 4.3.1 is satisfied for each $r = 1,2,\ldots$ with

$$H_i = \frac{K}{Y_i} \chi_{B_r^c}$$

and

$$h_i = \frac{\rho_i'}{y_i} \left(\frac{y_1 y_2}{\rho_1 y_1 + \rho_2 y_2}\right)^2 (\ell'(\Psi^{-1}(F)))^2 \chi_{B_r^c} \qquad i = 1 \text{ and } 2,$$

with $I = \{t: y_1(t) \wedge y_2(t) > 0\}$. Conditions II and III are also satisfied because with probability converging to 1 as $n \to \infty$, $|K| \le aK_C$ on $[t,\infty)$ for some fixed $a < \infty$ and $t < \sup I$; and the conditions of this proposition ensure that II and III are satisfied for the test statistic of COX (see Proposition 4.3.3, recalling that F is continuous). In the proof of Theorem 4.2.1, Conditions I, II and III and the fact that F is continuous, are used to show that the conditions of Theorem 2.4.1 are satisfied for each $r = 1,2,\ldots$ with $[0,\infty)$ replaced by $[0,\infty]$ and with

$$Z_i^{rn} = \int \frac{K}{Y_i} \chi_{B_r^c}\, dM_i \qquad , \quad i = 1,2,$$

$$\bar{Z}_i^{rn\varepsilon} = \int \frac{K}{Y_i} \chi_{B_r^c}\, \chi_{\left\{\left|\frac{K}{Y_1}\right| \vee \left|\frac{K}{Y_2}\right| > \varepsilon\right\}} dM_i, \quad i = 1,2;\ \varepsilon > 0,$$

and

$$A_i^r = \int \frac{\rho_{i'}}{y_i}\left(\frac{y_1 y_2}{\rho_1 y_1 + \rho_2 y_2}\right)^2 (\ell'(\Psi^{-1}(F)))^2 \chi_{B_r^c}\, dG, \quad i' \neq i = 1,2.$$

The conditions will also be satisfied for Z_i^n, $\bar{Z}_i^{n\varepsilon}$ and A_i defined by dropping the factor $\chi_{B_r^c}$ in the above three integrals provided that $A_i(\infty) < \infty$ and

$$(5.3.21) \qquad \lim_{r\to\infty} \limsup_{n\to\infty} P\left(\int_0^\infty \frac{K^2}{Y_i} \chi_{B_r}\, dG > \eta\right) = 0$$

for all $\eta > 0$ and each $i = 1,2$. The finiteness of $A_i(\infty)$ follows from (5.3.17) by the fact that $y_i dG \le dF$. Also (5.3.21) certainly holds if we remove (if ℓ' is unbounded) the interval $(0,t_r]$ from B_r for each r, because $\ell'(\Psi^{-1}(F))$ is bounded on the rest of B_r uniformly in r, and because by Proposition 4.3.3, (5.3.21) holds with K replaced by K_C. Condition (5.3.18) is equivalent to (5.3.21) with B_r replaced by $(0,t_r]$. So (5.3.21) holds in general. We have now established (5.3.19). By Lemma 4.3.1, for each r the analogous result to (5.3.20) with K replaced by $K\chi_{B_r^c}$ holds. But this result can be extended to the required one by using finiteness of $A_i(\infty)$ for each i, the relation (5.3.21), and Theorem 2.4.2 exactly as was done in the proof of Lemma 4.3.1 to make the extensions from I to $[0,u]$ and to $[0,\infty]$. □

<u>PROPOSITION 5.3.3</u>. *Under the combined conditions of Propositions* 5.3.1 *and* 5.3.2, *with* $\{F_\theta: \theta \in \Theta\}$ *given by* (5.3.13), $\log \frac{dP_1}{dP_0}$ *and* $W_{opt}(\infty)$ *are, under the null hypothesis, asymptotically bivariate normally distributed with a squared correlation coefficient equal to*

$$(5.3.22) \qquad \frac{\int_0^\infty \frac{y_1y_2}{\rho_1y_1+\rho_2y_2}(\ell'(\Psi^{-1}(F)))^2 dG}{\int_0^\infty (\rho_2y_1+\rho_1y_2)(\ell'(\Psi^{-1}(F)))^2 dG}.$$

(Under these conditions, (5.3.7) *can be replaced by* (5.3.16)*.)*

PROOF. For any real numbers α,β define

$$H_i^{\alpha\beta} = \pm\alpha 2\left(\sqrt{\frac{\lambda_{\theta_i^n}}{\lambda_{\theta_0}}} - 1\right) + \beta\ell'(\Psi^{-1}(\hat{F}_-))\frac{K_C}{Y_i} \qquad (\pm = (-1)^{i+1})$$

and

$$H_i^{\alpha\beta\varepsilon} = \pm\alpha 2\left(\sqrt{\frac{\lambda_{\theta_i^n}}{\lambda_{\theta_0}}} - 1\right)\chi_{\left\{2\left|\sqrt{\frac{\lambda_{\theta_i^n}}{\lambda_{\theta_0}}} - 1\right| \geq \frac{\varepsilon}{2\alpha}\right\}}$$

$$+ \beta\ell'(\Psi^{-1}(\hat{F}_-))\frac{K_C}{Y_i}\,\chi_{\left\{\ell'(\Psi^{-1}(\hat{F}_-))\frac{K^C}{Y_i} \geq \frac{\varepsilon}{2\beta}\right\}}.$$

For each (α,β) we shall verify the conditions of Theorem 2.4.1 with the interval $[0,\infty)$ replaced by $[0,\infty]$ and with

$$Z_i^{\alpha\beta n} = \int H_i^{\alpha\beta}\, dM_i \quad \text{in place of } Z_i^n,$$

$$\bar{Z}_i^{\alpha\beta n\varepsilon} = \int H_i^{\alpha\beta\varepsilon}\, dM_i \quad \text{in place of } \bar{Z}_i^{n\varepsilon}$$

and

$$A_i^{\alpha\beta} = \int \rho_{i'}\left(\alpha c + \frac{\beta}{y_i}\,\frac{y_1y_2}{\rho_1y_1+\rho_2y_2}\right)^2 (\ell'(\Psi^{-1}(F)))^2 y_i\, dG$$

$$\text{in place of } A_i \quad (i = 1,2).$$

After this, the Cramèr-Wold device gives the required result, with the asymptotic covariance of $\log \frac{dP_1}{dP_0}$ and $W(\infty)$ being equal to the coefficient of $2\alpha\beta$ in $A_1^{\alpha\beta}(\infty) + A_2^{\alpha\beta}(\infty)$. Now (in Propositions 5.3.1 and 5.3.2) we have already verified the conditions of Theorem 2.4.1 with $(\alpha,\beta) = (1,0)$ and $(\alpha,\beta) = (0,1)$. The condition involving $\langle \bar{Z}_i^{\alpha\beta n\varepsilon}, \bar{Z}_i^{\alpha\beta n\varepsilon}\rangle$ is now seen to hold for arbitrary (α,β) by writing

$$\langle \bar{Z}_i^{\alpha\beta n\varepsilon}, \bar{Z}_i^{\alpha\beta n\varepsilon} \rangle \leq 2\alpha^2 \langle \bar{Z}_i^{1\,0\,n\,\varepsilon/\alpha}, \bar{Z}_i^{1\,0\,n\,\varepsilon/\alpha} \rangle$$

$$+ 2\beta^2 \langle \bar{Z}_i^{0\,1\,n\,\varepsilon/\beta}, \bar{Z}_i^{0\,1\,n\,\varepsilon/\beta} \rangle .$$

It remains to show that

$$(5.3.23) \qquad \langle Z_i^{\alpha\beta n}, Z_i^{\alpha\beta n} \rangle (t) = \int_0^t (H_i^{\alpha\beta})^2 Y_i \, dG \to_P A_i^{\alpha\beta}(t)$$

as $n \to \infty$ for each $t \in [0,\infty]$ and $i = 1,2$. In fact we shall show

$$(5.3.24) \qquad \int_0^t \left(\alpha 2 n_i^{\frac{1}{2}} \left(\sqrt{\frac{\lambda_{\theta_i^n}}{\lambda_{\theta_0}}} - 1 \right) + \beta \ell'(\Psi^{-1}(\hat{F}_-)) \frac{n_i}{Y_i} n_i^{-\frac{1}{2}} K_C \chi_{B^c} \right)^2 \frac{Y_i}{n_i} dG$$

$$\to_P \int_0^t \rho_{i'} \left(\alpha c + \frac{\beta}{y_i} \frac{y_1 y_2}{\rho_1 y_1 + \rho_2 y_2} \chi_{B^c} \right)^2 (\ell'(\Psi^{-1}(F)))^2 y_i \, dG \qquad (i' \neq i),$$

as $n \to \infty$ for each $t \in [0,\infty]$ and for any $B \subset [0,\infty]$ such that $\ell'(\Psi^{-1}(F))$ is continuous and bounded outside B, and such that $B^c \subset [0,s]$ for some $s \in I$. After that we carry out the obvious extension procedure: we find a decreasing sequence of sets B'_r, each of which has the properties just required of B, such that $\bigcap_{r=1}^{\infty} B'_r$ equals the complement of I plus finitely many points, and such that

$$\lim_{r\to\infty} \limsup_{n\to\infty} P\left(\int_{B'_r} (\ell'(\Psi^{-1}(\hat{F}_-)))^2 K_C^2 Y_i^{-1} dG > \varepsilon \right) = 0$$

for all $\varepsilon > 0$. Then if (5.3.24) holds with $B = B'_r$ for each r, it holds with $B = \emptyset$; here we use the relation, for real functions f and g and a finite measure μ,

$$\left| \int_0^t (f + g\chi_{B^c})^2 d\mu - \int_0^t (f+g)^2 d\mu \right| = \left| \int_0^t \chi_B g^2 \, d\mu + 2\int_0^t \chi_B fg d\mu \right|$$

$$\leq \left| \int_B g^2 d\mu + 2\sqrt{\left(\left(\int_0^t f^2 d\mu \right) \left(\int_B g^2 d\mu \right) \right)} \right| .$$

Using the fact that Conditions II and III are satisfied for the test statistic of COX, we can take $B'_r = B_r \cup (s_r, \infty)$ for each r, where B_r is constructed in the proof of Proposition 5.3.2, and where $s_r = u$ for all r if $u \in I$, otherwise $s_r < u$ and $s_r \uparrow u$ as $r \to \infty$.

To return to the proof of (5.3.24), we recall from the proof of

Proposition 5.3.1 that

$$\pm 2n_i^{\frac{1}{2}}\left(\sqrt{\frac{\lambda_{\theta_i^n}}{\lambda_{\theta_0}}} - 1\right) = c\sqrt{\frac{n_{i'}}{n_1+n_2}}\,\frac{2}{\lambda_{\theta_0}^{\frac{1}{2}}}\,\frac{\lambda_{\theta_i^n}^{\frac{1}{2}}-\lambda_{\theta_0}^{\frac{1}{2}}}{\theta_i^n-\theta_0}$$

converges in $L^2(F)$ to

$$\rho_{i'}^{\frac{1}{2}}\, c\,\frac{\partial}{\partial\theta}\log\lambda_\theta\Big|_{\theta=\theta_0} = \rho_{i'}^{\frac{1}{2}}\, c\,\ell'(\Psi^{-1}(F)).$$

Also, by the properties of B,

$$\ell'(\Psi^{-1}(\hat{F}_-))\,\frac{n_i}{Y_i}\,n_i^{-\frac{1}{2}}\,K_C$$

converges uniformly on B^c to

$$\ell'(\Psi^{-1}(F))\,\frac{1}{Y_i}\,\rho_{i'}^{\frac{1}{2}}\,\frac{y_1 y_2}{\rho_1 y_1+\rho_2 y_2},$$

in probability, as $n\to\infty$. Since the latter function is bounded on B^c, the $L^2(F)$ distance between

$$\ell'(\Psi^{-1}(\hat{F}_-))\,\frac{n_i}{Y_i}\,n_i^{-\frac{1}{2}}\,K_C\chi_{B^c} \quad\text{and}\quad \ell'(\Psi^{-1}(F))\,\frac{\rho_{i'}^{\frac{1}{2}}}{Y_i}\,\frac{y_1 y_2}{\rho_1 y_1+\rho_2 y_2}\,\chi_{B^c}$$

converges in probability to zero as $n\to\infty$. Thus the difference, in $L^2(F)$, between

$$\alpha 2n_i^{\frac{1}{2}}\left(\sqrt{\frac{\lambda_{\theta_i^n}}{\lambda_{\theta_0}}} - 1\right) + \beta\ell'(\Psi^{-1}(\hat{F}_-))\,\frac{n_i}{Y_i}\,n_i^{-\frac{1}{2}}\,K_C\,\chi_{B^c}$$

and

$$\rho_{i'}^{\frac{1}{2}}\,\ell'(\Psi^{-1}(F))\left(\alpha c + \frac{\beta}{Y_i}\,\frac{y_1 y_2}{\rho_1 y_1+\rho_2 y_2}\,\chi_{B^c}\right)$$

converges in probability to zero as $n\to\infty$. Combining this fact with the two facts

$$\sup_{t\in(0,\infty)}\left|\frac{Y_i(t)}{n_i} - y_i(t)\right| \to_P 0 \quad\text{as } n\to\infty$$

and

$$P\left(\frac{Y_i}{n_i} \le \beta^{-1}(1-F)\text{ on }[0,\infty)\right) = 1 - o(1)$$

as $\beta \downarrow 0$ uniformly in n in the same way as was done in the proof of Proposition 5.3.1 yields (5.3.24). □

COROLLARY 5.3.1. *Suppose that the conditions of Propositions* 5.3.1 *and* 5.3.2 *hold, with* $\{F_\theta: \theta \in \Theta\}$ *given by* (5.3.13), *and with the asymptotic variances of* $\log \frac{dP_1}{dP_2}$ *and* $W_{opt}(\infty)$ *strictly positive. Then the efficiency of the best test of the class* K *(the one based on* K_{opt}*) with respect to the next powerful test for the sequence of alternatives is given by* (5.3.22). *This expression equals* 1 *if and only if* $y_1 = y_2$ *almost everywhere*-dF *where* $\ell'(\Psi^{-1}(F)) \neq 0$, *or for* $i = 1$ *or* 2, $\rho_i = 0$ *and* $y_i = 0$ *almost everywhere*-dF *where* $y_{i'} = 0$ *and* $\ell'(\Psi^{-1}(F)) \neq 0$ $(i' \neq i)$.

PROOF. That the efficiency is given by (5.3.22) is a straightforward application of Le Cam's third lemma. The conditions for an efficiency of 1 were investigated on page 113. □

As far as Examples 5.3.1 to 5.3.4 are concerned, the only difficulties in verifying the conditions of Corollary 5.3.1 occur with the verification of (5.3.18) for the case $\Psi = \Phi$, the standard normal distribution function. Now in this case, replacing $\hat{F}$ with $\tilde{F} = (n\hat{F}+1)/(n+1)$, we have

$$-\Phi^{-1}(\tilde{F}_-) \leq \ell'(\Psi^{-1}(\tilde{F}_-)) \leq -\Phi^{-1}(\tilde{F}_-) + 2\phi(0)$$

on $\{t: \tilde{F}_-(t) < \frac{1}{2}\}$. So in the presence of Conditions (5.2.4) and (5.2.5), (5.3.18) is equivalent to

$$\lim_{t\downarrow 0} \limsup_{n\to\infty} P\left(\int_0^t (\Phi^{-1}(\tilde{F}_-))^2 \chi_{\{\tilde{F}_- < \frac{1}{2}\}} dF > \varepsilon\right) = 0$$

under H_0 for all $\varepsilon > 0$. By (5.2.4), (5.2.5) and Proposition 3.2.1, this holds if

$$\lim_{t\downarrow 0} \int_0^t (\Phi^{-1}(\beta F))^2 dF = 0$$

for all $\beta > 0$. But by the change of variables $x = \Phi^{-1}(\beta F(t))$, the expression on the left hand side of this relation equals

$$\lim_{x\downarrow -\infty} \int_{-\infty}^x u^2 \frac{1}{\beta} \phi(u)du = 0$$

as required. This gives us

COROLLARY 5.3.2. *Under the conditions of Proposition* 4.3.3 *and with* F *continuous,* (5.3.19) *and* (5.3.20) *hold when* K_{opt} *is defined as in any of Examples* 5.3.1 *to* 5.3.4.

This result could have been extended to discontinuous F too, but we have not taken the trouble as it is hardly likely that one would use one of the new test statistics in such a case. Many authors indicate how asymptotically optimal test statistics might be constructed for the kind of situation we have considered; in particular PETO & PETO (1972), BROWN, HOLLANDER & KORWAR (1974), CROWLEY & THOMAS (1975) and PRENTICE (1978) all describe test statistics close to or identical to our proposal for the logistic distribution. However, as far as we know, no proof has been given that the hoped for properties of such test statistics do indeed hold in general.

The test statistics we constructed above were all members of $\mathcal{K}^+$. As examples of optimal test statistics for which K_{opt} is not nonnegative, we mention the case of varying shape parameters in the Weibull distribution, for which we obtain

$$\gamma \propto 1 - \log(-\log(1-F))$$

and the case of varying shape parameter σ in the lognormal distribution, for which

$$\gamma \propto \Phi^{-1}(F)(-\Phi^{-1}(F)-\phi(\Phi^{-1}(F))(1-F)^{-1}).$$

In each case, we suggest choosing the random weight function obtained by multiplying K_C with the above expressions after replacing the argument F with $\tilde{F}_-$.

5.4. Rényi-type tests

We have seen that test statistics in $\mathcal{K}$ can be constructed to have good properties when testing against particular parametric alternatives. At the same time, such test statistics will generally be consistent against alternatives of e.g. stochastic ordering (see Section 4.1). Still, it is conceivable that one would want consistency against the alternative of mere inequality of F_1 and F_2. In this section we show how this can be (nearly) attained by means of a simple modification of the test statistics in $\mathcal{K}$, while retaining some of the good power properties against special alternatives.

We consider asymptotic behaviour under a fixed null and a fixed alternative hypothesis; i.e. either $F_1^n = F_2^n = F$ for all n (H_0) or $F_1^n = F_1$ and $F_2^n = F_2$ for all n, $F_1 \neq F_2$ (H_1). Suppose as usual that (5.2.4) and (5.2.5) hold, where unlike the case of contiguous alternatives, the functions y_1 and y_2 will generally depend on whether one is working under H_0 or H_1. Let $u \in (0,\infty)$ be fixed and satisfy $y_1(u) > 0$ and $y_2(u) > 0$ both under H_0 and H_1. Now consider a test statistic in $\mathcal{K}$ for which

$$\sqrt{\frac{n_1+n_2}{n_1n_2}}\, K$$

converges uniformly on $[0,u]$ to a function k under H_0 and H_1 (again, the function k will generally depend on whether one is working under H_0 or H_1). Suppose in each case that k is left continuous with right hand limits and k_+ of bounded variation on $[0,u]$. Applying Theorem 4.2.1 and Lemma 4.3.1, it follows that under H_0, as $n \to \infty$,

$$W = \int K\left(\frac{dN_1}{Y_1} - \frac{dN_2}{Y_2}\right) \to_{\mathcal{D}} Z_0^\infty \quad \text{in } D[0,u],$$

where Z_0^∞ is a zero mean Gaussian process with independent increments and variance function

$$\operatorname{var}(Z_0^\infty(t)) = \sum_{i=1}^{2} \int_0^t \frac{\rho_{i'}}{y_i} k^2(1-\Delta G)\,dG \qquad (i' \neq i);$$

also

$$V_\ell(u) \to_P \operatorname{var}(Z_0^\infty(u)), \quad \ell = 1 \text{ or } 2.$$

On the other hand, under H_1, as $n \to \infty$,

$$\int K\left(\frac{dN_1}{Y_1} - \frac{dN_2}{Y_2}\right) - \int K(dG_1 - dG_2) \to_{\mathcal{D}} Z_1^\infty \quad \text{in } D[0,u],$$

where Z_1^∞ has the same properties as Z_0^∞ except that its variance function is now given by

$$\operatorname{var}(Z_1^\infty(t)) = \sum_{i=1}^{2} \int_0^t \frac{\rho_{i'}}{y_i} k^2(1-\Delta G_i)\,dG_i \qquad (i' \neq i);$$

also

$$V_1(u) \to_P \operatorname{var}(Z_1^\infty(u)),$$

$$V_2(u) \to_P \sum_{i=1}^{2} \int_0^t \frac{\rho_i}{y_{i'}} k^2\left(1 - \frac{\rho_1 y_1 \Delta G_1 + \rho_2 y_2 \Delta G_2}{\rho_1 y_1 + \rho_2 y_2}\right) dG_i \qquad (i' \neq i),$$

and

$$\sup_{t\in[0,u]} \left| \sqrt{\frac{n_1+n_2}{n_1 n_2}} \int_0^t K(dG_1 - dG_2) - \int_0^t k(dG_1 - dG_2) \right| \to_P 0.$$

(For the result on $V_\ell(u)$ see Section 5.1, especially formulae (5.1.1) and (5.1.2).) Now choosing $\ell = 1$ or 2 suppose that the limit in probability of $V_\ell(u)$ is strictly positive under H_0 and H_1. Then arguing as on page 80, we see that under H_0, as $n \to \infty$,

$$U = \frac{\sup_{t\in[0,u]} |W(t)|}{\sqrt{V_\ell(u)}} \to_{\mathcal{D}} \sup_{t\in A} |B(t)| \leq \sup_{t\in[0,1]} |B(t)|, \tag{5.4.1}$$

where B is a standard continuous Brownian motion on [0,1] and $A \subset [0,1]$ is the range of the function $\operatorname{var}(Z_0^\infty(\cdot))/\operatorname{var}(Z_0^\infty(u)) : [0,u] \to [0,1]$. So $A = [0,1]$ if F is continuous.

However under H_1, as $n \to \infty$,

$$U \to_P \infty$$

unless $\int k(dG_1-dG_2)$ is identically zero on $[0,u]$. This can only happen if, under H_1, $k = 0$ on $[0,u]$ almost everywhere-$d\mu$ where $\frac{dG_1}{d\mu} \neq \frac{dG_2}{d\mu}$, where μ is a σ-finite measure dominating G_1 and G_2. In particular, if under H_1 k is positive on $[0,t]$ for some $t \leq u$ such that F_1 and F_2 differ on $[0,t]$, then the test of H_0 based on the test statistic U is consistent against H_1. Note that if we base the test on the distribution of $\sup_{t\in[0,1]} |B(t)|$ even if $A \neq [0,1]$, it becomes a conservative test.

More information is given on this distribution on page 81. The two-sample procedure we have proposed here can be considered as an extension of the one-sample confidence-band technique we discussed in Section 4.2, which itself extended a method of RÉNYI (1953); hence our name "Rényi-type tests". It can also be considered as a Kolmogorov-Smirnov type test, since it is based on the maximum distance between two empirical processes, here $\int K \frac{dN_1}{Y_1}$ and $\int K \frac{dN_2}{Y_2}$. A related class of test statistics is described by

FLEMING & HARRINGTON (1980), whose work is also based on AALEN (1976). KOZIOL & PETKAU (1978) propose the test statistic U in the special case when $K = K_C$ (corresponding to the test statistic of COX) and when the censoring is simple Type II (Example 3.1.2).

It is interesting to compare the test statistic U with its natural competitor

$$U' = \frac{|W(u)|}{\sqrt{V_\ell(u)}}$$

(where the same ℓ has been chosen as in the definition of U). It is not possible to standardize U in some fixed way so as to obtain an equivalent test statistic, asymptotically normally distributed with fixed variance both under the null hypothesis and under a contiguous alternative hypothesis. So if a comparison between U and U' is to be made in terms of Pitman asymptotic relative efficiency of U with respect to U', care is needed in defining this concept in the first place. Defining it as the limit, for a sequence of alternatives approaching the null hypothesis, of the ratio of the sample sizes required by size α tests based on U' and U respectively to achieve power β at each alternative in the sequence, it will depend on α and β. However, a theorem of WIEAND (1976) gives conditions under which this asymptotic efficiency has a limit as $\alpha \downarrow 0$ independent of $\beta \in (0,1)$.

Application of WIEAND's theorem shows that in one very general case of interest, and under suitable regularity conditions, the limiting Pitman efficiency of U with respect to U' equals 1. This is the case of the ordered hazard type of alternative hypothesis - $dG_1 \le dG_2$ on $[0,u]$ or $dG_1 \ge dG_2$ on $[0,u]$ - and of a random weight function K whose limiting weight function k is positive on $[0,u]$. The explanation of this result is that in this situation, the two quantities

$$\sup_{t\in[0,u]} \left|\int_0^t k(dG_1 - dG_2)\right| \quad \text{and} \quad \left|\int_0^u k(dG_1 - dG_2)\right|,$$

which play an important role in determining the asymptotic behaviour under a fixed alternative of U and U' respectively, are equal; while the tail behaviour of the limiting null hypothesis distributions of U and U' respectively is the same too. However more attention needs to be paid to the small sample properties of the test statistic U before too much weight is attached to this result.

CHAPTER 6

GENERAL CENSORSHIP AND TRUNCATION

In previous chapters we have only considered so-called right censored observations of n lifetimes $X_1,\ldots,X_n$. Furthermore we have supposed that in a natural time scale each lifetime starts at time zero; in other words, at *time* t each object still under observation has *age* t. In Examples 3.1.1, 3.1.2 and 3.1.5, the experiment being modelled already had this property; in Example 3.1.4 on the other hand independence between the observations was used to realign the X_i's without causing any problems.

In this chapter we shall informally discuss a model for censored observations $X_1,\ldots,X_n$ in which we allow the time of birth to be different for each object; we also allow for far more general schemes of partial observation of these lifetimes than previously. For simplicity we restrict attention to the one-sample case in which $X_1,\ldots,X_n$ are independent and identically distributed with a distribution function F which we want to estimate. Finally we shall illustrate our remarks by looking again at Example 3.1.6. For other examples we refer to HYDE (1977) and LAGAKOS, SOMMER & ZELEN (1978). Our approach is similar to HYDE's (1977).

For convenience we shall take as usual as time axis the positive half line $[0,\infty)$. Let $T_1,\ldots,T_n \geq 0$ be n random *birth times*, and let $X_1,\ldots,X_n$ be the corresponding n lifetimes; we suppose that $X_1,\ldots,X_n$ are independent and identically distributed with distribution function F satisfying $F(0) = 0$. We say that object i is born at time T_i and dies at time $T_i + X_i$.

However this system is only partially observed. We suppose that there also exist n random *observation processes* $J_1,\ldots,J_n$ defined on $[0,\infty)$ and taking values in $\{0,1\}$ such that if $J_i(t) = 1$ then object i is alive and under observation just before time t; in this case we suppose that we know the object's *age* $t - T_i$ and can observe whether or not it dies at this moment; i.e. whether or not $t - T_i = X_i$. In particular it follows that J_i is zero outside the time interval $(T_i,T_i+X_i]$. If in the interval $(T_i,T_i+X_i]$ the sample paths of J_i are nonincreasing and left continuous, partial observation of the i-th lifetime results in a censored lifetime $\tilde{X}_i$ and an indicator random variable δ_i such that $\delta_i = 1 \Rightarrow \tilde{X}_1 = X_i$, $\delta_i = 0 \Rightarrow \tilde{X}_i < X_i$. However we shall not make this restriction in this chapter.

We shall have to make some kind of assumption concerning the possible dependence between the observation processes $J_1,\ldots,J_n$ and the lifetimes $X_1,\ldots,X_n$. As in Section 3.1 we wish to exclude the possibility of statistical dependence between whether or not an object has been or is being observed and its remaining lifetime. We shall formulate such an assumption by imitating Assumptions 3.1.1 and 3.1.2, for which we shall assume that $X_1,\ldots,X_n$, $T_1,\ldots,T_n$, $J_1,\ldots,J_n$ are defined on some stochastic basis $(\Omega,F,P),\{F_t\colon t \in [0,\infty)\}$. We also define for each $i = 1,\ldots,n$ and each $t \in [0,\infty)$

(6.1) $$N_i(t) = \chi_{\{T_i+X_i\le t,\, J_i(X_i+T_i)=1\}}$$

(6.2) $$L_i(t) = (t-T_i)\chi_{[T_i,\infty)}(t)$$

(6.3) $$M_i(t) = N_i(t) - \int_0^t J_i(s)dG(L_i(s)), \quad \text{where } G = \int (1-F)^{-1}dF$$

(recall that J_i is zero outside $(T_i,T_i+X_i]$).

Our assumptions then become:

<u>ASSUMPTION 6.1</u>. With respect to the stochastic basis $(\Omega,F,P),\{F_t\colon t\in[0,\infty)\}$, for each $i = 1,\ldots,n$, T_i and $T_i + X_i$ are stopping times, J_i is a predictable process and M_i is a square integrable martingale with

$$\langle M_i,M_i\rangle = \int J_i(1 - \Delta G(L_i))dG(L_i)$$

and

$$\langle M_i,M_{i'}\rangle = 0 \quad (i' \ne i).$$

<u>ASSUMPTION 6.2</u>. For each t, conditional on F_{t-}, $\Delta N_1(t),\ldots,\Delta N_n(t)$ are independent zero-one random variables with expectations $J_1(t)\Delta G(L_2(t)),\ldots, J_n(t)\Delta G(L_n(t))$.

Even though the censoring is more general, the new assumptions can be interpreted exactly as Assumptions 3.1.1 and 3.1.2 were; the only difference is that the lifetime of the n objects start at times $T_1,\ldots,T_n$ instead of time zero. Note that the process N_i counts 1 at the death of object i if and when death is observed. Thus if F has a continuous hazard rate λ, we are stating that given what has happened up to time t, the probability of observing the death of object i in the time interval $[t,t+h]$ is zero if $J_i(t) = 0$; otherwise it is approximately $h\lambda(t-T_i)$ where $t-T_i$ is the object's current age.

If for each i, $T_i = 0$ almost surely and J_i has the properties described above leading to right censored observations, Assumptions 6.1 and 6.2 are equivalent to 3.1.1 and 3.1.2.

What can be observed are the processes J_i, and for each i and t such that $J_i(t) = 1$, the age of object i at time t and whether or not death occurs at that time instant. To estimate F we shall first want to pool our observations, and this leads us to define for $s \in [0,\infty)$

$$(6.4) \qquad N(s) = \#\{i: X_i \leq s, J_i(T_i+X_i) = 1\}$$

$$(6.5) \qquad Y(s) = \#\{i: J_i(T_i+s) = 1\}.$$

Here the argument s refers to *age*: N(s) is the number of deaths observed at an age $\leq$ s, and Y(s) is the number of objects which were under observation at age s. It is again natural to estimate F with the product limit estimator defined with respect to N and Y_i, i.e. by

$$(6.6) \qquad \hat{F}(t) = 1 - \prod_{s\leq t} \frac{\Delta N(s)}{Y(s)} .$$

However it is not clear whether $\hat{F}$ will have the same properties as we established for it in Chapters 3 and 4.

In the special case $T_1 = \ldots = T_n = 0$ almost surely, we can easily generalize the old results. (Such a model is also discussed by AALEN (1976) with the further restriction that F should have a hazard rate.) Defining

$$(6.7) \qquad M = N - \int YdG$$

we have in this case $N = \sum_{i=1}^n N_i$, $Y = \sum_{i=1}^n J_i$, and $M = \sum_{i=1}^n M_i$, so that M is a square integrable martingale with $\langle M,M\rangle = \int Y(1-\Delta G)dG$ and Y is a predictable process. Also for each t, conditional on $\mathcal{F}_{t-}$, $\Delta N(t)$ is binomially distributed with parameters Y(t), $\Delta G(t)$. In deriving results on the product limit estimator in Chapters 3 and 4, the only further properties of N and Y we used were some of the properties of the paths of Y: left continuous and nondecreasing. These properties no longer hold and proofs will have to be modified accordingly. For instance in Theorem 4.1.1 the condition "$Y(t) \to_P \infty$" would have to be replaced by "$\sup_{s\in[0,t]} Y(s) \to_P \infty$".

If we cannot suppose that $T_1 = \ldots = T_n = 0$, the process M defined by (6.4), (6.5) and (6.7) is not necessarily a martingale. However we shall show that it still has the same mean and covariance structure, and indicate the significance of this result. Define for each age s and time t

$$H_i^s(t) = J_i(s)\chi_{[0,s]}(L_i(t)).$$

It is easy to verify that

$$N(s) = \sum_{i=1}^{n} \int_0^\infty H_i^s(t)dN_i(t).$$

This suggests we also evaluate

$$\sum_{i=1}^{n} \int_0^\infty H_i^s(t)J_i(t)dG(L_i(t))$$

$$= \sum_{i=1}^{n} \int_{t\in(T_i,T_i+s]} J_i(t)dG(t-T_i)$$

$$= \int_0^s Y(u)dG(u).$$

Thus

$$\sum_{i=1}^{n} \int_0^\infty H_i^s dM_i = N(s) - \int_0^s YdG = M(s).$$

But for given s, H_i^s is a bounded predictable process and therefore by (2.2.1) and Assumption 6.1,

$$EM = E(N - \int YdG) = 0, \tag{6.8}$$

or equivalently,

$$EN = \int EYdG.$$

Similarly using (2.2.2) we obtain

$$(6.9) \qquad E(M(s)M(s')) = E\Big(\Big(\sum_{i=1}^{n}\int_0^\infty H_i^s dM_i\Big)\Big(\sum_{i=1}^{n}\int_0^\infty H_i^{s'} dM_i\Big)\Big)$$

$$= E\sum_{i=1}^{n}\int_0^\infty H_i^s H_i^{s'} J_i(1-\Delta G(L_i))dG(L_i)$$

$$= E\sum_{i=1}^{n}\int_0^\infty H_i^{s\wedge s'} J_i(1-\Delta G(L_i))dG(L_i)$$

$$= \int_0^{s\wedge s'} EY(1-\Delta G)dG.$$

Thus although M is perhaps not a square integrable martingale with $\langle M,M\rangle =$ $= \int Y(1-\Delta G)dG$, it has exactly the same mean and covariance structure as if it were. This fact, together with the representation (3.2.13) of $(\hat{F}-F)/(1-F)$ as an integral with respect to M, suggests that *if*, as in Theorem 4.2.2, convergence in probability of Y/n implies convergence in distribution of $n^{\frac{1}{2}}(\hat{F}-F)$, then the limiting distribution of $n^{\frac{1}{2}}(\hat{F}-F)$ will be of the same form as in Theorem 4.2.2 and we will be able to base asymptotic confidence band procedures on the observable processes N and Y exactly as was done after Theorem 4.2.2.

Before illustrating this point further, let us mention a useful extension of the above model. We have assumed that at most n lifetimes could have been observed. However there are no real difficulties involved in allowing the total number of lifetimes specified in the model to be infinite (so that we specify lifetimes $X_1,X_2,\ldots$, birth times $T_1,T_2,\ldots$ and observation processes $J_1,J_2,\ldots$). We still define N, Y, $\hat{F}$ and M by (6.4) to (6.7), and as long as $E(N(\infty)) < \infty$ we can establish (6.8) and (6.9) by monotone convergence and L^2 convergence respectively. The censoring implied by the J_i's is really a mixture of censoring and truncation: objects i for which the realized path of J_i is identically zero are not registered by the processes N and Y and one does not even have to know which or how many objects are of this kind.

With this last extension we can finally discuss Example 3.1.6. First we consider a single replacement sequence; i.e. we start with a single object and replace it at death with a new one, and continue till a fixed length of time u has elapsed. Thus we let $X_1,X_2,\ldots$ be the independent and identically distributed lifetimes, we define the birth times by $T_1 = 0$ and $T_n = \sum_{i=1}^{n-1} X_i$, $n = 2,3,\ldots$, and define the observation processes J_i by $J_i(t) = 1 \Leftrightarrow T_{i-1} < t \le T_i\wedge u$. Assumptions 6.1 and 6.2 are easily verified for the natural choice of F_t using some of the counting process theory of Section 2.3. Example 3.1.6 is concerned with n independent copies of this model.

Two different asymptotic approaches are now available; let u become large or let n become large. The case $u \to \infty$ is of course rather trivial as far as this specific model is concerned. However more general replacement models in which objects can be replaced before death lead to great difficulties and so far no general results are known. BATHER (1977) describes such a model in which a death is more costly than a planned replacement. As time evolves an estimate of F and the corresponding cost minimizing replacement policy are improved.

In the case $n \to \infty$ the results suggested above do hold (see GILL (1978, 1980)). Of course we can no longer apply a martingale central limit theorem to $n^{\frac{1}{2}}(\hat{F} - F)/(1 - F)$, but the independence between the n copies allows us to apply the weak law of large numbers to Y/n and the central limit theorem to $n^{-\frac{1}{2}}M$, and (3.2.13) links these to $n^{\frac{1}{2}}(\hat{F} - F)$.

Appendix 1

Proof of Theorem 2.3.1.

Here we exploit the properties of the so-called *optional quadratic variation* process $[M,M]$ associated with a local martingale M (see MEYER (1976) or JACOD (1979)).

Consider first the case $r = 1$ and drop the index i. N is locally bounded, and by the proof of MEYER (1976) Theorem IV.12, so is A. Since $[M,M](t) = \sum_{s\leq t} \Delta M(s)^2$ it turns out by expanding $(\Delta M(s))^2$ that

$$[M,M] = \int (1 - 2\Delta A)\, dM + \int (1 - \Delta A)\, dA.$$

$1 - 2\Delta A$ is a locally bounded predictable process and M is a local martingale, hence $\int (1 - 2\Delta A)\, dM$ is a local martingale, with paths of locally bounded variation. Since the processes $1 - \Delta A$ and A are predictable, so is $\int (1 - \Delta A)\, dA$; and of course it too has paths of locally bounded variation. Combining these facts and using MEYER (1976) Chapter IV, we see that $[M,M]$ is locally bounded and hence locally integrable. This implies that M is a local square integrable martingale. In this case, $\langle M,M\rangle$ is equal to the dual predictable projection of $[M,M]$; so

$$\langle M,M\rangle = \int (1 - \Delta A)\, dA.$$

Since the paths of $\langle M,M\rangle$ are non-decreasing, we now see that $0 \leq \Delta A \leq 1$. So $1 - 2\Delta A$ is a bounded predictable process. If T is a stopping time such that $E N(T) < \infty$, then $E A(T) < \infty$, and M^T (the process M stopped at T) is a martingale of integrable variation. Consequently $E \int_0^T (1 - 2\Delta A)\, dM = 0$; also $E \int_0^T (1 - \Delta A)\, dA < \infty$; and so $E[M,M](T) < \infty$. But for any local martingale M, $E[M,M](T) < \infty$ implies that M^T is a square integrable martingale.

Now we consider the case $r > 1$. All that remains to be proved is that $\langle M_i, M_j\rangle = -\int \Delta A_i\, dA_j$. If $i \neq j$, $N_i + N_j$ is also a counting process, whose compensator must be $A_i + A_j$. So

$$\langle M_i + M_j, M_i + M_j\rangle = \int (1 - \Delta A_i - \Delta A_j)(dA_i + dA_j),$$

while by bilinearity and symmetry of $\langle \cdot, \cdot \rangle$,

$$\langle M_i + M_j, M_i + M_j\rangle = \langle M_i, M_i\rangle + \langle M_j, M_j\rangle + 2\langle M_i, M_j\rangle.$$

Combining gives the required result. □

Appendix 2

On constructing a stochastic basis

If σ-algebras F_t are defined in some natural way, as in formula (2.3.6) or in the statement of Theorem 3.1.2, it is not immediately obvious that they form a stochastic basis: in particular, it is not obvious that $\{F_t: t \in [0,\infty)\}$ is right continuous. Here we give a theorem of DE SAM LAZARO (1974) which answers these and related questions in a very general setting. First we need some notation and definitions.

Let (Ω, F, P) be an arbitrary probability space, and let $(Z, \mathcal{Z})$ be an arbitrary measurable space. A Z-valued function x on $[0,\infty)$ is called a *jump function* if for each $t \in [0,\infty)$ an $\varepsilon > 0$ exists such that x is constant on $[t, t+\varepsilon]$. A process $X = \{X(t,\omega): t \in [0,\infty), \omega \in \Omega\}$ is called a *jump process* if for each t, $X(t)$ is a measurable mapping from (Ω, F) to $(Z, \mathcal{Z})$, and if for *each* ω, the sample path $X(\cdot,\omega)$ is a jump function on $[0,\infty)$ with values in Z.

<u>THEOREM A.2.1</u>. *Let* X *be a jump process, and define*

$$F_t^0 = \sigma\{X(s): s \le t\}.$$

Then $\{F_t^0: t \in [0,\infty)\}$ *is right continuous. Furthermore, if* T *is any* $\{F_t^0\}$ *stopping time, then*

$$F_T^0 = \sigma\{X(s \wedge T): s \in [0,\infty)\}.$$

<u>PROOF</u>. See DE SAM LAZARO (1974) Lemma 3.3. This proof is elegant and elementary, and can be read independently from the rest of the paper if one notes that in it, the reference to the first part of Proposition 3.1 should be to the second part of Proposition 2.1. □

<u>COROLLARY A.2.1</u>. *Let* X *be a jump process, and let* A *be an arbitrary sub* σ*-algebra of* F. *Define*

$$F_t = A \vee \sigma\{X(s): s \le t\}.$$

Then $\{F_t\}$ *is right continuous, and if* T *is any* $\{F_t\}$ *stopping time,*

$$F_T = A \vee \sigma\{X(s \wedge T): s \in [0,\infty)\} = A \vee \sigma\{T, X(s \wedge T): s \in [0,\infty)\}. \tag{A.2.1}$$

PROOF. Define a jump process $\tilde{X}$ with values in the measurable space $(Z\times\Omega, \mathcal{Z}\otimes\mathcal{A})$ by

$$\tilde{X}(t,\omega) = (X(t,\omega),\omega).$$

Since $\mathcal{F}_t = \sigma\{\tilde{X}(s): s \le t\}$ and $\mathcal{A} \vee \sigma\{X(s\wedge T): s \in [0,\infty)\} = \sigma\{\tilde{X}(s\wedge T): s \in [0,\infty)\}$ the result is immediate (T can be included in the final expression of (A.2.1) since it is automatically $\mathcal{F}_T$ measurable). □

From Corollary A.2.1, we see that if $\mathcal{F}$ is complete and $\mathcal{A}$ contains all P-null sets of $\mathcal{F}$, then $(\Omega,\mathcal{F},P),\{\mathcal{F}_t: t \in [0,\infty)\}$ forms a stochastic basis.

In a typical application of Theorem A.2.1, we might be given a probability space $(\Omega,\mathcal{F},P)$, on which are defined random time instants (i.e. $[0,\infty]$-valued random variables) $T_1,\ldots,T_k$, and a further k random variables $Y_1,\ldots,Y_k$ which are supposed to be "realised" or become observable at the time instants $T_1,\ldots,T_k$. We wish to construct σ-algebras $\mathcal{F}_t$ relative to which $T_1,\ldots,T_k$ are stopping times and which reflect the availability of Y_i from time T_i. This can be done via the construction of a jump process X with values in $\mathbb{R}^{2k}$, defined by

$$X(t) = ((\chi_{\{T_i\le t\}}, Y_i\cdot\chi_{\{T_i\le t\}}): i = 1,\ldots,k).$$

We then get

$$\mathcal{F}_t^0 = \sigma\{X(s): s \le t\} = \sigma\{(\chi_{\{T_i\le t\}}, T_i\chi_{\{T_i\le t\}}, Y_i\chi_{\{T_i\le t\}}): i=1,\ldots,k\}$$

and

$$\mathcal{F}_t = \mathcal{F}_t^0 \vee \mathcal{A},$$

where $\mathcal{A}$ is the set of all P-null sets of $\mathcal{F}$ (supposed to be complete) and their complements. So defined, $(\Omega,\mathcal{F},P),\{\mathcal{F}_t: t \in [0,\infty)\}$ is a stochastic basis; $T_1,\ldots,T_k$ are stopping times; and for *any* stopping time T,

$$\mathcal{F}_T = \mathcal{A} \vee \sigma\{T,(\chi_{\{T_i\le T\}}, T_i\chi_{\{T_i\le T\}}, Y_i\chi_{\{T_i\le T\}}): i = 1,\ldots,k\}.$$

(In fact T itself can be omitted from the list of generating random variables, but the above form is easier to interpret.)

The same construction works for random time instants T_α, $\alpha \in A$, with an arbitrary index set A, provided that for all $\omega \in \Omega$, for every $t \in [0,\infty)$ an $\varepsilon > 0$ exists such that for all $\alpha \in A$, $T_\alpha(\omega) \notin (t,t+\varepsilon]$. If this property only holds for P-almost all $\omega \in \Omega$, then the construction can be applied provided that the T_α's are first redefined on the exceptional set. After that, augmenting $F_t^{0^\alpha}$ with all P-null sets of F as above yields a stochastic basis, which in fact does not depend on how the T_α's have been modified.

Appendix 3

Proof of Theorem 2.3.4

Following JACOD (1975,1979), the stochastic bases constructed in the course of the following proof do not necessarily satisfy the completeness assumption ((iii) on page 8).

By altering N on a null set of $\mathcal{F}$, we may suppose that all the paths of N are nondecreasing, right continuous, zero at time zero, and integer-valued with jumps of size +1 only. We may redefine $T_0, T_1, \ldots$ accordingly; and we can alter A on a null set of $\mathcal{F}$ so that all of its paths are zero at time zero and satisfy

$$t \in (T_n, T_{n+1}] \Rightarrow A(t) - A(T_n) = f_n(t-T_n; T_1, \ldots, T_n).$$

By the completeness of $\{\mathcal{F}_t: t \in [0,\infty)\}$, A and N remain adapted processes after this alteration. (It is not immediate that A is still predictable, but we do not need this fact anyway.) Next, define σ-algebras $\mathcal{F}_t^N$, $t \in [0,\infty]$, by

$$\mathcal{F}_t^N = \sigma\{N(s): s \leq t\}.$$

$(\Omega, \mathcal{F}_\infty^N, P), \{\mathcal{F}_t^N: t \in [0,\infty)\}$ forms a stochastic basis on which N is a counting process, all of whose paths have the usual properties. By JACOD (1979) Proposition 3.39, A is a predictable process with respect to this new stochastic basis; and all its paths are nondecreasing, right continuous, and zero at time zero. It is also easy to verify that N - A remains a martingale; so A is still the compensator of N.

Let X be the set of nondecreasing, right continuous, integer-valued functions on $[0,\infty)$ which are zero at time zero and make jumps of size +1 only. Letting $x = \{x_t: t \in [0,\infty)\}$ denote the generic member of X, define σ-algebras on X by

$$\mathcal{X}_t = \sigma\{x_s: s \leq t\}, \quad t \in [0,\infty].$$

Define on $(X, \mathcal{X}_\infty)$ measurable functions $\tau_n = \inf\{t: x_t \geq n\}$, $n = 0,1,\ldots$; and define a process $a = \{a_t: t \in [0,\infty)\}$ on $(X, \mathcal{X}_\infty)$ by

$$a_0 = 0 \quad \text{and} \quad t \in (\tau_n, \tau_{n+1}] \Rightarrow a_t - a_{\tau_n} = f_n(t-\tau_n; \tau_1, \ldots, \tau_n).$$

Finally define a probability measure P^N on $(X,\mathcal{X}_\infty)$ by

$$P^N = P\circ\phi^{-1},$$

where ϕ is the measurable mapping

$$\phi\colon (\Omega,\mathcal{F}^N_\infty) \to (X,\mathcal{X}_\infty),$$

defined by $\phi(\omega) = N(\cdot,\omega)$. We now see that

$$(X,\mathcal{X}_\infty,P^N),\{\mathcal{X}_t\colon t \in [0,\infty)\}$$

is a stochastic basis, on which χ is a counting process and (by JACOD (1979) Proposition 3.39 again) a is a predictable process. a has right continuous, nondecreasing paths, zero at time zero. Also for all $t \in [0,\infty)$, $\mathcal{F}^N_t = \phi^{-1}(\mathcal{X}_t)$, and by definition $P^N = P\circ\phi^{-1}$. Therefore by JACOD (1979) Theorem 10.37, $\chi - a$ is a martingale, so a is the compensator of χ.

Had we started off with a different stochastic basis, and a different counting process N', satisfying the conditions of the theorem with the *same* functions $f_0,f_1,\ldots$, we would have proved that a is also the compensator of χ with respect to the stochastic basis $(X,\mathcal{X}_\infty,P^{N'}),\{\mathcal{X}_t\colon t \in [0,\infty)\}$. Therefor by JACOD (1975) Theorem 3.4, P^N and $P^{N'}$ coincide on $\mathcal{X}_\infty$. But the joint probability distributions of $T_1,T_2,\ldots$ and $T_1',T_2',\ldots$ can be recovered from P^N and $P^{N'}$ respectively, and the theorem is proved. □

Appendix 4

Proof of Lemma 3.2.1

We shall derive Lemma 3.2.1 as a corollary to the following proposition:

PROPOSITION A.4.1. *Let* A *and* B *be right continuous nondecreasing functions on* $[0,\infty)$, *zero at time zero; suppose* $\Delta A \le 1$ *and* $\Delta B < 1$ *on* $[0,\infty)$. *Then the unique locally bounded solution* Z *of*

$$Z(t) = \int_{s\in[0,t]} \frac{1 - Z(s-)}{1 - \Delta B(s)}(dA(s) - dB(s)) \tag{A.4.1}$$

is given by

$$Z(t) = 1 - \frac{\prod_{s\le t}(1-\Delta A(s))\exp(-A_c(t))}{\prod_{s\le t}(1-\Delta B(s))\exp(-B_c(t))}, \tag{A.4.2}$$

where it should be recalled that A_c *is the continuous part of* A, *defined by*

$$A_c(t) = A(t) - \sum_{s\le t} \Delta A(s). \tag{A.4.3}$$

PROOF. We adapt the proof of LIPTSER & SHIRYAYEV (1978) Lemma 18.8, which deals with the case where B is identically zero. We shall make use of the following simple results: if U and V are right continuous functions of locally bounded variation on $[0,\infty)$, then for all $t \in [0,\infty)$

$$U(t)V(t) = U(0)V(0) + \int_{s\in(0,t]} U(1-)dV(s) + \int_{s\in(0,t]} V(s)dU(s), \tag{A.4.4}$$

which can also be written in the form

$$d(UV) = U_-dV + VdU. \tag{A.4.5}$$

From this one can easily derive

$$d(U^r) = \left(\sum_{i=0}^{r-1} U^i U_-^{r-1-i}\right)dU, \qquad r = 1,2,\ldots \tag{A.4.6}$$

and

$$d(U^{-1}) = -(UU_-)^{-1}\, dU. \tag{A.4.7}$$

If U is nondecreasing and nonnegative, then (A.4.6) gives

$$rU_-^{r-1}dU \le d(U^r) \le rU^{r-1}dU, \quad r = 1,2,\ldots . \tag{A.4.8}$$

Let us first show that (A.4.2) does define a solution to (A.4.1). It is certainly locally bounded. Define

$$U(t) = \prod_{s\le t} \frac{1 - \Delta A(s)}{1 - \Delta B(s)}$$

and

$$V(t) = \exp(-A_c(t) + B_c(t)).$$

Then if (A.4.2) holds,

$$\begin{aligned}
Z(t) = 1 - U(t)V(t) &= 1 - U(0)V(0) - \int_{s\in(0,t]} U(s-)dV(s) - \int_{s\in(0,t]} V(s)dU(s) \\
&= - \int_{s\in(0,t]} U(s-)V(s)(-dA_c(s) + dB_c(s)) \\
&\quad - \sum_{s\le t} V(s)U(s-)\left(\frac{1 - \Delta A(s)}{1 - \Delta B(s)} - 1\right) \\
&= \int_{s\in[0,t]} \frac{1 - Z(s-)}{1 - \Delta B(s)}(dA_c(s) - dB_c(s)) \\
&\quad + \sum_{s\le t} \frac{1 - Z(s-)}{1 - \Delta B(s)}(\Delta A(s) - \Delta B(s)) \\
&= \int_{s\in[0,t]} \frac{1 - Z(s-)}{1 - \Delta B(s)}(dA(s) - dB(s)),
\end{aligned}$$

where $(1-\Delta B)^{-1}$ could be introduced into the integrand because A_c and B_c are continuous.

Next, suppose Z' is another locally bounded solution of (A.4.1). Define $\tilde{Z} = Z-Z'$, $L(t) = \sup|\tilde{Z}(s)|$, $\alpha = \int (1-\Delta B)^{-1}(dA+dB)$. Then for any $s \le t$

$$|\tilde{Z}(s)| \le \int_{u\in[0,s]} |\tilde{Z}(u-)|d\alpha(u) \le L(t)\alpha(s).$$

Substituting the outer inequality back in the first one gives

$$|\tilde{Z}(s)| \le \int_{u\in[0,s]} L(t)\alpha(u-)d\alpha(u) \le \tfrac{1}{2}L(t)\alpha(s)^2$$

by (A.4.8) with r = 2. Repeating this procedure, we find that for any r,

$$|\tilde{Z}(s)| \leq \frac{L(t)}{r!}\alpha(s)^r \to 0 \quad \text{as} \quad r \to \infty. \qquad \square$$

COROLLARY: Proof of Lemma 3.2.1.

(3.2.9) holds for t such that $G(t) < \infty$ by setting $B = 0$ and $A = G$ in (A.4.1). If $G(t) \uparrow \infty$ as $t \uparrow \tau$ for some $\sigma > 0$, then (3.2.9) must also hold for $t = \tau$ by taking limits. Since $G = \int (1 - F_-)^{-1} dF$, in this case we must have $F(t) \uparrow 1$ as $t \uparrow \sigma$, and so $\sigma = \tau$ and (3.2.9) holds for all $t > \tau$.

We have now proved assertion (i). The only non-trivial part of (ii) is to show that $F(t) \uparrow 1$ as $t \uparrow \tau$ implies $G(t) \uparrow \infty$ as $t \uparrow \tau$. Now for each $t < \tau$, $\sup_{s \in [0,t]} \Delta G(s) < 1$. By (3.2.9), taking logarithms and carrying out a Taylor expansion,

$$-G(t) - \frac{1}{2}C(t)G(t) \leq -G(t) - \frac{1}{2}C(t) \sum_{s \leq t} \Delta G(s)^2 \leq \log(1 - F(t)),$$

where

$$C(t) = \sup_{s \in [0,t]} (1 - \Delta G(s))^{-1} < \infty$$

for each $t < \tau$. If $F(t) \uparrow 1$ as $t \uparrow \tau$ then either $G(t) \uparrow \infty$ or $\limsup_{t \uparrow \tau} \Delta G(t) = 1$; but the latter equality also implies that $G(t) \uparrow \infty$.

Assertion (iii) follows immediately from (i) since continuity of F implies continuity of G.

Finally by (3.2.6) and (3.2.9) for t such that $F(t) < 1$, putting $A = \int \frac{dN}{Y}$ and $B = G$ in (A.4.2) shows that

$$Z = 1 - \frac{1 - \hat{F}}{1 - F}$$

solves (A.4.1) with the present choice of A and B. But with this Z, A, and B, (A.4.1) is equivalent to (3.2.12) by the equality $(1 - F(s-))(1 - \Delta G(s)) = 1 - F(s)$. $\square$

Appendix 5

Asymptotic normality of an estimator of mean lifetime

Many authors consider estimation of mean lifetime $\int_0^\infty t dF(t) = \int_0^\infty (1-F(t))dt$ on the basis of the product limit estimator. However either no attempt at proof is made (KAPLAN & MEIER (1958), BRESLOW & CROWLEY (1974)), or boundedness assumptions are made: YANG (1977) assumes that $F(t) = 1$ for some $t < \infty$ and FLEMING (1978) only considers estimation of $\int_0^t (1-F(s))ds$ for some t such that $F(t) < 1$. (In these two cases Theorem 4.2.3 and Theorem 4.2.2 respectively can be applied directly.) The estimator considered is always $\int_0^T t\, d\hat{F}(t)$ or $\int_0^T (1-\hat{F}(t))dt$ where $T = \max_j \tilde{X}_j$ (the notation here is as in the second part of Section 4.2). These quantities are related by

$$\int_0^T (1-\hat{F}(t))dt = \int_0^T t\, d\hat{F}(t) + T(1-\hat{F}(T)).$$

Here we shall consider $\hat{\mu}_T = \int_0^T (1-\hat{F}(t))dt$ and define a corresponding function μ by

$$\mu_t = \int_0^t (1-F(s))ds$$

and process $\hat{\mu}$ by

$$\hat{\mu}_t = \int_0^t (1-\hat{F}(s)) \frac{1-F(s)}{1-F^T(s)} ds$$

where $F^T(s) = F(s \wedge T)$. We also define a function $\bar{\mu}$ by

$$\bar{\mu}_t = \int_t^\infty (1-F(s))ds.$$

We assume throughout that $\mu_\infty = \bar{\mu}_0 < \infty$.

We shall give conditions for asymptotic normality of $n^{\frac{1}{2}}(\hat{\mu}_T-\mu_T)$; consistency of $\hat{\mu}_T$ was mentioned on page 58. We shall assume that $F(t) < 1$ for all $t < \infty$, $F(\infty) = 1$ and $T \to_P \infty$ as $n \to \infty$. We shall not give conditions for $n^{\frac{1}{2}}(\mu_\infty-\mu_T) = n^{\frac{1}{2}}\bar{\mu}_T \to_P 0$ as $n \to \infty$, though we shall mention an example where it holds.

Before stating our theorem, let us note one application of our results which is not so obvious: namely to the Total Time on Test Plot of BARLOW & CAMPO (1975). This is a plot of an estimate of

$$\int_0^{F^{-1}(p)} (1-F(s))ds \Big/ \int_0^{\infty} (1-F(s))ds$$

against $p \in [0,1]$. We propose that for censored data the plot should be made with $\hat{F}$ instead of F and T instead of ∞ in this formula (BARLOW & CAMPO (1975) suggest the use of $N/N(\infty)$ rather than $\hat{F}$), so our results give conditions for the denominator here to behave respectably.

THEOREM A.5.1. *Assume the conditions hold given in the first sentence of Theorem 4.2.3 and define* y, $\tilde{\tau}$ *and* u *as was done there. Suppose furthermore that* $u = \infty$ *(so that* $T \to_P \infty$ *as* $n \to \infty$*) and* $F(u) = 1$. *Then under the conditions*

$$\lim_{t\uparrow\infty} \bar{\mu}_t^{-2} \int_0^t ((1-F)(1-F_-)(1-L_-))^{-1}\, dF = 0 \tag{A.5.1}$$

and

$$\lim_{t\uparrow\infty} \limsup_{n\to\infty} \int_t^{\infty} \bar{\mu}^{-2}((1-F)(1-F_-)(1-L_-^n))^{-1} \chi_{(0,1]}(L_-^n)\, dF = 0 \tag{A.5.2}$$

we have

$$n^{\frac{1}{2}}(\hat{\mu}_T - \mu_T) \to_{\mathcal{D}} N(0,\sigma^2) \quad \text{as } n \to \infty, \tag{A.5.3}$$

where

$$\sigma^2 = \int_0^{\infty} \bar{\mu}^{-2}((1-F)(1-F_-)(1-L_-))^{-1}\, dF < \infty. \tag{A.5.4}$$

σ^2 *can be consistently estimated by*

$$\int_0^T \left(\int_t^T (1-\hat{F}(s))ds\right)^2 \frac{n\chi_{\{Y>1\}}}{Y-1} \frac{dN}{Y}.$$

PROOF. Let Z be defined as in the proof of Theorem 4.2.3. We have

$$\begin{aligned} n^{\frac{1}{2}}(\hat{\mu}-\mu) &= n^{\frac{1}{2}}\left(\int (1-\hat{F}) \frac{1-F}{1-F^T} ds - \int (1-F)ds\right) \\ &= -\int n^{\frac{1}{2}} \frac{\hat{F}-F^T}{1-F^T}(1-F)ds = \int Z d\bar{\mu} \\ &= \bar{\mu}Z - \int \bar{\mu} dZ. \end{aligned}$$

To prove (A.5.3) for some σ^2 it suffices to show that for all $\varepsilon > 0$

$$\lim_{t\uparrow\infty} \limsup_{n\to\infty} P\left(\sup_{s\in(t,\infty)} \left|\left(\bar{\mu}Z - \int\bar{\mu}dZ\right)(s) - \left(\bar{\mu}Z - \int\bar{\mu}dZ\right)(t)\right| > \varepsilon\right) = 0$$

and that the same holds with $\limsup_{n\to\infty}$ omitted and with Z^{∞} instead of Z. We can consider the parts $\bar{\mu}Z$ and $\int \bar{\mu}dZ$ separately. Now the second part is easy to deal with in the usual way since it is a square integrable martingale with predictable variation process

$$\int \bar{\mu}^{2}\left(\frac{1-\hat{F}_{-}}{1-F}\right)^{2} n \frac{J}{Y} (1-\Delta G)\,dG.$$

We use the inequality of LENGLART (Theorem 2.4.2), in which we bound $1-\hat{F}_{-}$ with $\beta^{-1}(1-F_{-})$ and $n\,J/Y$ with $\beta^{-1}((1-F_{-})(1-L_{-}^{n}))^{-1}\chi_{(0,1]}(L_{-}^{n})$ according to Theorem 3.2.1 and VAN ZUIJLEN (1978) Theorem 1.1 and Corollary 3.1 respectively. The part $\bar{\mu}Z$ can be dealt with exactly as was $(1-F)Z$ in Theorem 4.2.3. Running through the proof of that theorem we see that (A.5.1) and (A.5.2) correspond to (4.2.2) and (4.2.3); each time a term $(1-F)^{2}$ has been replaced by $\bar{\mu}^{2}$.

This proves weak convergence of the process $n^{\frac{1}{2}}(\hat{\mu}-\mu)$ in $D[0,\infty]$. Since obviously $T \to_{P} \infty$ as $n \to \infty$ we also have asymptotic normality of $n^{\frac{1}{2}}(\hat{\mu}_{T}-\mu_{T})$. By the proof we have $\lim_{t\to\infty} \bar{\mu}_{t} Z^{\infty}(t) = 0$ almost surely and so the limiting variance of $n^{\frac{1}{2}}(\hat{\mu}_{T}-\mu_{T})$ has no component corresponding to $\bar{\mu}Z$ and thus is given by (A.5.4) corresponding to $\int \bar{\mu}dZ$ only. Consistency of the estimator of this variance follows by similar arguments to those used in the proof of Theorem 4.2.3, noting also the remarks on consistency of $\hat{\mu}_{T}$ on page 58. □

Note that

$$\bar{\mu}^{2} \int ((1-F)(1-F_{-})(1-L_{-}))^{-1}\,dF =$$

$$\int \bar{\mu}^{2}((1-F)(1-F_{-})(1-L_{-}))^{-1}dF + \int \left(\int ((1-F)(1-F_{-})(1-L_{-}))^{-1}dF\right)d(\bar{\mu}^{2})$$

so that (A.5.2) implies that the limit in (A.5.1) exists, but not necessarily that it is zero (cf. the remarks after Theorem 4.2.3).

In the case of no censoring, these conditions become

$$\lim_{t\uparrow\infty} \bar{\mu}_{t}^{2} \frac{F(t)}{1-F(t)} = 0$$

and

$$\int_{0}^{\infty} \bar{\mu}_{t}^{2}\, d\left(\frac{F(t)}{1-F(t)}\right) < \infty$$

since by (A.4.7)

$$((1-F)(1-F_-))^{-1}\,dF = d(\tfrac{F}{1-F}) = d(1-(1-F)^{-1}).$$

Now

$$\bar{\mu}_t^2 \frac{F(t)}{1-F(t)} = \int_0^t \bar{\mu}_s^2\, d(\frac{F(s)}{1-F(s)})$$
$$+ 2\int_0^t (1-(1-F(s))^{-1})\bar{\mu}(s)d\bar{\mu}(s)$$
$$= \int_0^t \bar{\mu}_s^2\, d(\frac{F(s)}{1-F(s)}) - 2\int \bar{\mu}_s\, ds + \bar{\mu}_t^2 - \bar{\mu}_0^2.$$

We have

$$2\int_0^t \bar{\mu}_s\, d_s = 2\int_0^t \left(\int_s^\infty\left(\int_u^\infty dF(v)\right)du\right)ds$$
$$= 2\int_0^t \left(\int_0^v\left(\int_0^u ds\right)du\right)dF(v) = \int_0^t v^2\, dF(v).$$

Thus

$$\int_0^t \bar{\mu}_s^2\, d(\frac{F(s)}{1-F(s)}) = \int_0^t v^2 dF(v) - (\bar{\mu}_0^2-\bar{\mu}_t^2) + \bar{\mu}_t^2\frac{F(t)}{1-F(t)},$$

and conditions (A.5.1) and (A.5.2) are in the case of no censoring equivalent to

$$\text{(A.5.5)} \qquad \operatorname{var}(X_j) < \infty \quad \text{and} \quad \lim_{t\to\infty} \bar{\mu}_t^2 \frac{F(t)}{1-F(t)} = 0.$$

Now

$$\frac{\bar{\mu}_t^2\, F(t)}{1-F(t)} = (E(X_j-t\,|\,X_j > t))^2\cdot F(t)(1-F(t))$$

so a slightly stronger assumption is

$$\operatorname{var}(X_j) < \infty \quad \text{and} \quad \limsup_{t\to\infty} E(X_j-t\,|\,X_j > t) < \infty.$$

This certainly holds when F has an increasing hazard rate which is the case for many realistic limetime models (e.g. gamma distribution, Weibull distribution with shape parameter ≥ 1, exponential distribution, normal distribution. The lognormal distribution does not have an increasing hazard rate but (A.5.5) is satisfied for it too).

As a second example consider the case of an exponential distribution with exponentially distributed censoring, $1-F(t) = e^{-t}$ and $1-L^n(t) = e^{-\beta t}$ for all n, so that β represents the degree of censoring. It is now easy to check that (A.5.1) and (A.5.2) hold if and only if $\beta < 1$. In this case $\bar{\mu}_t = e^{-t}$ and T tends to infinity like log n so that $n^{\frac{1}{2}} \bar{\mu}_T \to_P 0$ as $n \to \infty$, and we have

$$n^{\frac{1}{2}}(\hat{\mu}_T - \mu_\infty) \to_{\mathcal{D}} \mathcal{N}(0,\sigma^2)$$

as $n \to \infty$.

Appendix 6

Proof of a theorem of Daniels

Here we sketch a proof inspired by TAKÁCS (1967) though our argument is geometric rather than combinatorial.

THEOREM A.6.1 (DANIELS (1945), ROBBINS (1954)). *Let* $\hat{F}$ *be the empirical distribution function based on a random sample of size* n *from the continuous distribution function* F. *Then*

$$P(\hat{F}(t) \leq \beta^{-1} F(t)\ \forall t) = 1-\beta \quad \forall \beta \in [0,1].$$

PROOF. It suffices to consider the case when F is the uniform distribution on [0,1]. Extend indefinitely and repetitively the graph of $\hat{F}$ and of $\beta^{-1}F$ as in Figure A.6.1. We imagine the extended graph of $\hat{F}$ as a staircase or mountain side, on which the sun shines with rays parallel to the line $\beta^{-1}F$. The probability required is the probability that at O the sun can be seen, or alternatively 1 minus the probability that O is in shadow.

Let $Y_1 < \ldots < Y_n$ be the order statistics of the random sample and define $Y_{n+r} = 1 + Y_r$, $r = 1,\ldots,n$. Let R be a random variable uniformly distributed on $\{1,\ldots,n\}$ independently of the sample, and condition on the horizontal step lengths $Z_1 = Y_{R+1} - Y_R, \ldots, Z_n = Y_{R+n} - Y_{R+n-1}$ (i.e. we forget that it is a step of length $Y_{n+1} - Y_n$ on which O lies, and condition only on the shape of the staircase). It is easy to see that conditional on these lengths, the point O lies uniformly distributed on the horizontal sections $Z_1,\ldots,Z_n$ ($\sum_{i=1}^n Z_i = 1$). Now of these sections a length exactly $1-\beta$ is in the light and β is in shadow (see Figure A.6.1; there *are* points in the light even if O is not). Thus conditional on $Z_1,\ldots,Z_n$ the required probability is $1-\beta$ and unconditionally it must be too. □

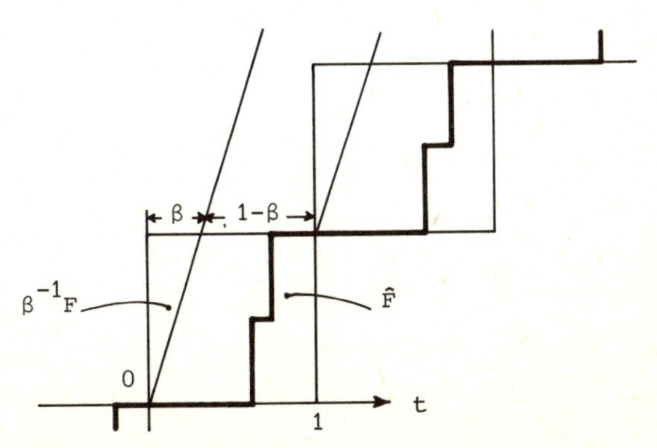

Figure A.6.1.

REFERENCES

AALEN, O.O. (1976), *Statistical Theory for a Family of Counting Processes*, Inst. of Math. Stat., Univ. of Copenhagen, Copenhagen.

AALEN, O.O. (1977), *Weak Convergence of Stochastic Integrals related to Counting Processes*, Z. Wahrscheinlichkeitstheorie und verw. Gebiete 38, p.261-277.

AALEN, O.O. (1978), *Nonparametric Inference for a Family of Counting Processes*, Ann. Statist. 6, p.701-726.

AALEN, O.O. & S. JOHANSEN (1978), *An Empirical Transition Matrix for Nonhomogeneous Markov Chains based on Censored Observations*, Scand. J. Statist. 5, p.141-150.

BARLOW, R.E. & R. CAMPO (1975), *Total Time on Test Processes and Applications to Failure Data Analysis*, p.451-481 in: Reliability and Fault Tree Analysis, R.E. Barlow, J.F. Fussel & N.D. Singpurwalla (eds), SIAM, Philadelphia.

BATHER, J.A. (1977), *On the Sequential Construction of an Optimal Age Replacement Policy*, Bull. Int. Stat. Inst. 47, p.253-266.

BETHLEHEM, J.G., DOES, R.J.M.M. & R.D. GILL (1977), *Verdelingsvrije Methoden bij Censurering*, Report SN 6, Dept. of Math. Stat., Mathematisch Centrum, Amsterdam.

BILLINGSLEY, P. (1968), *Weak Convergence of Probability Measures*, Wiley, New York.

BIRNBAUM, Z.W. & A.W. MARSHALL (1961), *Some Multivariate Chebyshev Inequalities with Extensions to Continuous Parameter Processes*, Ann. Math. Statist. 32, p.687-703.

BOEL, R., VARAIYA, P. & E. WONG (1975a), *Martingales on Jump Processes, I: Representation Results*, SIAM J. Control 13, p.999-1021.

BOEL, R., VARAIYA, P. & E. WONG (1975b), *Martingales on Jump Processes, II: Applications*, SIAM J. Control 13, p.1022-1061.

BRÉMAUD, P. (1975), *The Martingale Theory of Point Processes over the Real Half Line admitting an Intensity*, p. 519-542 in: Control Theory, Numerical Methods and Computer System Modelling,

A. Bensoussan & J.L. Lions (eds), Lecture Notes in Economics and Mathematical Systems 107, Springer-Verlag, Berlin.

BRÉMAUD, P. & J. JACOD (1977), *Processus Ponctuels et Martingales: Résultats Récents sur la Modelisation et le Filtrage,* Adv. Appl. Prob. 9, p.362-416.

BRESLOW, N. (1970), *A Generalized Kruskal-Wallis Test for Comparing K Samples Subject to Unequal Patterns of Censorship,* Biometrika 57, p.579-594.

BRESLOW, N. (1974), *Covariance Analysis of Censored Survival Data,* Biometrics 30, p.89-99.

BRESLOW, N. & J. CROWLEY (1974), *A Large Sample Study of the Life Table and Product Limit Estimates under Random Censorship,* Ann. Statist. 2, p.437-453.

BRESLOW, N. (1975), *Analysis of Survival Data under the Proportional Hazards Model,* Int. Stat. Rev. 43, p.45-58.

BROWN, B.W., HOLLANDER, M. & R.M. KORWAR (1974), *Nonparametric Tests for Independence with Censored Data, with Applications to Heart Transplant Studies,* p.327-354 in: Reliability and Biometry, F. Proschan & R.J. Serfling (eds), SIAM, Philadelphia.

COX, D.R. (1972), *Regression Models and Life-tables,* J. Roy. Statist. Soc. B. 34, p.187-200 (with discussion).

COX, D.R. (1975), *Partial Likelihood,* Biometrika 62, p.269-276.

CROWLEY, J. & D.R. THOMAS (1975), *Large Sample Theory for the Log Rank Test,* Technical Report no. 415, Dept. of Statist., University of Wisconsin, Madison, Wisconsin.

DANIELS, H.E. (1945), *The Statistical Theory of the Strength of Bundles of Threads, I,* Proc. Roy. Soc. A 183, p.405-435.

DOLIVO, F.G. (1974), *Counting Processes and Integrated Conditional Rates: a Martingale Approach with Application to Detection Theory,* Ph.D. thesis, University of Michigan.

DUDLEY, R.M. (1968), *Distances of Probability Measures and Random Variables,* Ann. Math. Statist. 39, p.1563-1572.

EFRON, B. (1967), *The Two Sample Problem with Censored Data,* Proc. Fifth Berkeley Symp. Math. Stat. Prob. 4, p.831-853.

ELLIOT, R.J. (1976), *Stochastic Integrals for Martingales of a Jump Process with Partially Acessible Jump Times*, Z. Wahrscheinlichkeitstheorie und verw. Gebiete 36, p.213-226.

FELLER, W. (1968), *An Introduction to Probability Theory and its Applications*, Vol. I (3rd Edition), Wiley, New York.

FELLER, W. (1971), *An Introduction to Probability Theory and its Applications*, Vol. II (2nd Edition), Wiley, New York.

FLEMING, T.R. & D.P. HARRINGTON (1980), *A Class of Hypothesis Tests for One and Two Sample Censored Survival Data*, Dept. of Appl. Math. and Comp. Sci. Report 80-9, University of Virginia.

FÖLDES, A., REJTŐ, L. & B.B. WINTER (1980), *Strong Consistency Properties of Nonparametric Estimators for Randomly Censored Data, I: The Product-Limit Estimator; II: Estimation of Density and Failure Rate* (to appear in Periodica Math. Hung.).

FÖLDES, A. & L. REJTŐ (1980a), *Asymptotic Properties of the Nonparametric Survival Curve Estimators under Variable Censoring* (to appear in Proceedings of the PSMS Symposium, Lecture Notes in Mathematics, Springer-Verlag, Berlin).

FÖLDES, A. & L. REJTŐ (1980b), *Strong Uniform Consistency for Nonparametric Survival Curve Estimators from Randomly Censored Data* (to appear in Ann. Statist.).

GEHAN, E.A. (1965), *A Generalized Wilcoxon Test for Comparing Arbitrarily Singly-Censored Samples*, Biometrika 52, p.203-223.

GILL, R.D. (1978), *Testing with Replacement and the Product Limit Estimator*, Report SW 57, Dept. of Math. Stat., Mathematisch Centrum, Amsterdam (condensed version to appear in Ann. Statist.)

GILL, R.D. (1980), *Nonparametric Estimation based on Censored Observations of a Markov Renewal Process*, Z. Wahrscheinlichkeitstheorie und verw. Gebiete 53, p.97-116.

GILLESPIE, M.J. & L. FISHER (1979), *Confidence Bands for the Kaplan-Meier Survival Curve Estimate*, Ann. Statist. 7, p.920-924.

HÁJEK, J. & Z. ŠIDÁK (1967), *Theory of Rank Tests*, Academic Press, New York.

HALL, W.J. & J.A. WELLNER (1980), *Confidence Bands for a Survival Curve from Censored Data*, Biometrika 67, p.133-143.

HELLAND, I.S. (1980), *Central Limit Theorems for Martingales with Discrete or Continuous Time*, submitted to Adv. Appl. Probability.

HOEFFDING, W. (1956), *On the Distribution of the Number of Successes in Independent Trials*, Ann. Math. Statist. 27, p.713-721.

HOLLANDER, M. & F. PROSCHAN (1979), *Testing to Determine the Underlying Distribution using Randomly Censored Data*, Biometrics 35, p.393-401.

HYDE, J. (1977), *Testing Survival under Right Censoring and Truncation*, Biometrika 64, p.225-230.

JACOD, J. (1973), *On the Stochastic Intensity of a Random Point Process over the Half-Line*, Technical Report 51, Dept. of Stat., Princeton Univ.

JACOD, J. (1975), *Multivariate Point Processes: Predictable Projection, Radon-Nikodym Derivatives, Representation of Martingales*, Z. Wahrscheinlichkeitstheorie und verw. Gebiete 31, p.235-253.

JACOD, J. (1979), *Calcul Stochastique et Problèmes de Martingales*, Lecture Notes in Mathematics 714, Springer-Verlag, Berlin.

JOHANSEN, S. (1978), *The Product Limit Estimator as Maximum Likelihood Estimator*, Scand. J. Statist. 5, p.195-199.

KALBFLEISCH, J.D. & R.L. PRENTICE (1973), *Marginal Likelihoods based on Cox's Regression and Life Models*, Biometrika 60, p.267-278.

KAPLAN, E.L. & P. MEIER (1958), *Nonparametric Estimation from Incomplete Observations*, J. Amer. Statist. Assoc. 53, p.457-481.

KOZIOL, J.A. & A.J. PETKAU (1978), *Sequential Testing of the Equality of Two Survival Distributions using the Modified Savage Statistic*, Biometrika 65, p.615-623.

LAGAKOS, S.W., SOMMER, C.J. & M. ZELEN (1978), *Semi Markov Models for Censored Data*, Biometrika 65, p.311-317.

LE CAM, L. (1960), *Locally Asymptotically Normal Families of Distributions*, University of California Publications in Statistics 3, p.37-98.

LENGLART, E. (1977), *Relation de Domination entre deux Processus*, Ann. Inst. Henri Poincaré 13, p.171-179.

LINDVALL, T. (1973), *Weak Convergence of Probability Measures and Random Functions in the Function Space* D[0,∞), J. Appl. Prob. 10, p.109-121.

LIPTSER, R.S. & A.N. SHIRYAYEV (1978), *Statistics of Random Processes, II: Applications*, Springer-Verlag, Berlin.

LIPTSER, R.S. & A.N. SHIRYAYEV (1980), *A Functional Central Limit Theorem for Semimartingales*, to appear.

MANTEL, N. (1966), *Evaluations of Survival Data and Two New Rank Order Statistics Arising in its Consideration*, Cancer Chemother. Rep. 50, p.163-170.

MANTEL, N. (1967), *Ranking Procedures for Arbitrarily Restricted Observation*, Biometrics 23, p.65-78.

MEIER, P. (1975), *Estimation of a Distribution Function from Incomplete Observations*, p.67-87 in: Perspectives in Probability and Statistics, J. Gani (ed.), Applied Prob. Trust, Sheffield.

MEYER, P.A. (1971), *Demonstration Simplifiée d'un Théorème de Knight*, p.191-195 in: Séminaire de Probabilités V, Lecture Notes in Mathematics 191, Springer-Verlag, Berlin.

MEYER, P.A. (1972), *Temps d'Arrêt Algébriquement Prévisibles*, p.159-163 in: Séminaire de Probabilités VI, Lecture Notes in Mathematics 258, Springer-Verlag, Berlin.

MEYER, P.A. (1976), *Un Cours sur les Intégrales Stochastiques*, p.245-400 in: Séminaire de Probabilités X, Lecture Notes in Mathematics 511, Springer-Verlag, Berlin.

MURALI-RAO, K. (1969), *On Decomposition Theorems of Meyer*, Math. Scand. 24, p.66-78.

NELSON, W. (1972), *Theory and Applications of Hazard Plotting for Censored Failure Data*, Technometrics 14, p.945-966.

PETERSON, A.V. (1975), *Nonparametric Estimation in the Competing Risks Problem*, Technical Report no. 73, Dept. of Statistics, Stanford University, Stanford.

PETERSON, A.V. (1977), *Expressing the Kaplan-Meier Estimator as a Function of Empirical Subsurvival Functions*, J. Amer. Statist. Assoc. 72, p.854-858.

PETO, R. (1972), *Rank Tests of Maximal Power against Lehmann-type Alternatives*, Biometrika 59, p.472-474.

PETO, R. & J. PETO (1972), *Asymptotically Efficient Rank Invariant Test Procedures*, J.R. Statist. Soc. (A) 135, p.185-206.

PRENTICE, R.L. (1978), *Linear Rank Tests with Right Censored Data*, Biometrika 65, p.167-179.

PURI, M.L. & P.K. SEN (1971), *Nonparametric Methods in Multivariate Analysis*, Wiley, New York.

RAO, U.V.R., SAVAGE, I.R. & M. SOBEL (1960), *Contributions to the Theory of Rank Order Statistics: Two Sample Censored Case*, Ann. Math. Statist. 31, p.415-426.

REBOLLEDO, R. (1978), *Sur les Applications de la Théorie des Martingales à l'Étude Statistique d'une Famille de Processus Ponctuels*, p.27-70 in: Journées de Statistique des Processus Stochastiques (Proceedings, Grenoble 1977), Lecture Notes in Mathematics 636, Springer-Verlag, Berlin.

REBOLLEDO, R. (1979a), *Central Limit Theorems for Local Martingales*, to appear in Z. Wahrscheinlichkeitstheorie und verw. Gebiete.

REBOLLEDO, R. (1979b), *Décomposition de Martingales Locales et Raréfaction des Sauts*, p.138-146 in: Séminaire de Probabilités XIII, Lecture Notes in Mathematics 721, Springer-Verlag, Berlin.

REBOLLEDO, R. (1979c), *La Méthode de Martingales Appliquée à l'Étude de la Convergence en Loi de Processus*, to appear in Mémoires du Soc. Math. de France.

RÉNYI, A. (1953), *On the Theory of Order Statistics*, Acta Math. Acad. Sci. Hungar. 4, p.191-231.

RÉNYI, A. (1963), *On the Distribution Function* L(z), Selected Translations in Math. Statist. and Probability 4, p.219-224.

ROBBINS, H. (1954), *A One-Sided Confidence Interval for an Unknown Distribution Function*, Ann. Math. Statist. 25, p.409.

DE SAM LAZARO, J. (1974), *Sur les Hélices du Flot Spécial sous une Fonction,* Z. Wahrscheinlichkeitstheorie und verw. Gebiete 30, p.279-302.

SAVAGE, I.R. (1956), *Contributions to the Theory of Rank Order Statistics - the Two-Sample Case,* Ann. Math. Stat. 27, p.590-615.

SHORACK, G.R. & J.A. WELLNER (1978), *Linear Bounds on the Empirical Distribution Function,* Ann. Probab. 6, p.349-353.

STONE, C. (1963), *Weak Convergence of Stochastic Processes defined on Semi-infinite Time Intervals,* Proc. Am. Math. Soc. 14, p.694-696.

TAKÁCS, L. (1967), *Combinatorial Methods in the Theory of Stochastic Processes,* Wiley, New York.

THOMAS, D.R. (1969), *Conditionally Locally Most Powerful Rank Tests for the Two-Sample Problem with Arbitrarily Censored Data,* Technical Report no. 7, Dept. of Statistics, Oregon State University.

THOMAS, D.R. (1975), *On a Generalized Savage Statistic for Comparing Two Arbitrarily Censored Samples,* Technical Report, Dept. of Statistics, Oregon State University.

TSIATIS, A.A. (1978), *An Example of Nonidentifiability in Competing Risks,* Scand. Actuarial J. 1978, p.235-239.

VERVAAT, W. (1972), *Success Epochs in Bernoulli Trials (with Applications in Number Theory),* Mathematical Centre Tracts 42, Mathematisch Centrum, Amsterdam.

WALSH, J.E. (1962), *Handbook of Nonparametric Statistics (Vol. I: Investigation of Randomness, Moments, Percentiles and Distributions),* Van Nostrand, Princeton.

WICHURA, M.J. (1970), *On the Construction of Almost Uniformly Convergent Random Variables with Given Weakly Convergent Laws,* Ann. Math. Statist. 41, p.284-291.

WIEAND, H.S. (1974), *On a Condition under which the Pitman and Bahadur Approaches to Efficiency Coincide,* Ph.D. dissertation, Univ. of Maryland.

WIEAND, H.S. (1976), *A Condition under which Pitman and Bahadur Approaches to Efficiency Coincide,* Ann. Statist. 4, p.1003-1011.

WINTER, B.B., FÖLDES, A. & L. REJTŐ (1978), *Glivenko-Cantelli Theorems for the Product Limit Estimate,* Problems of Control and Information Theory 7, p.213-225.

YANG, G. (1977), *Life Expectancy under Random Censorship,* Stochastic Processes and their Applications 6, p.33-39.

VAN ZUIJLEN, M.C.A. (1977), *Empirical Distributions and Rank Statistics,* Mathematical Centre Tracts 79, Mathematisch Centrum, Amsterdam.

VAN ZUIJLEN, M.C.A. (1978), *Properties of the Empirical Distribution Function for Independent Nonidentically Distributed Random Variables,* Ann. Probability 6, p.250-266.

SUBJECT INDEX

AUTHOR INDEX

OTHER TITLES IN THE SERIES MATHEMATICAL CENTRE TRACTS

A leaflet containing an order-form and abstracts of all publications mentioned below is available at the Mathematisch Centrum, Kruislaan 413, Amsterdam 1098SJ, The Netherlands. Orders should be sent to the same address.

MCT 1 T. VAN DER WALT, *Fixed and almost fixed points*, 1963. ISBN 90 6196 002 9.

MCT 2 A.R. BLOEMENA, *Sampling from a graph*, 1964. ISBN 90 6196 003 7.

MCT 3 G. DE LEVE, *Generalized Markovian decision processes, part I: Model and method*, 1964. ISBN 90 6196 004 5.

MCT 4 G. DE LEVE, *Generalized Markovian decision processes, part II: Probabilistic background*, 1964. ISBN 90 6196 005 3.

MCT 5 G. DE LEVE, H.C. TIJMS & P.J. WEEDA, *Generalized Markovian decision processes, Applications*, 1970. ISBN 90 6196 051 7.

MCT 6 M.A. MAURICE, *Compact ordered spaces*, 1964. ISBN 90 6196 006 1.

MCT 7 W.R. VAN ZWET, *Convex transformations of random variables*, 1964. ISBN 90 6196 007 X.

MCT 8 J.A. ZONNEVELD, *Automatic numerical integration*, 1964. ISBN 90 6196 008 8.

MCT 9 P.C. BAAYEN, *Universal morphisms*, 1964. ISBN 90 6196 009 6.

MCT 10 E.M. DE JAGER, *Applications of distributions in mathematical physics*, 1964. ISBN 90 6196 010 X.

MCT 11 A.B. PAALMAN-DE MIRANDA, *Topological semigroups*, 1964. ISBN 90 6196 011 8.

MCT 12 J.A.TH.M. VAN BERCKEL, H. BRANDT CORSTIUS, R.J. MOKKEN & A. VAN WIJNGAARDEN, *Formal properties of newspaper Dutch*, 1965. ISBN 90 6196 013 4.

MCT 13 H.A. LAUWERIER, *Asymptotic expansions*, 1966, out of print; replaced by MCT 54 and 67.

MCT 14 H.A. LAUWERIER, *Calculus of variations in mathematical physics*, 1966. ISBN 90 6196 020 7.

MCT 15 R. DOORNBOS, *Slippage tests*, 1966. ISBN 90 6196 021 5.

MCT 16 J.W. DE BAKKER, *Formal definition of programming languages with an application to the definition of ALGOL 60*, 1967. ISBN 90 6196 022 3.

MCT 17 R.P. VAN DE RIET, *Formula manipulation in ALGOL 60, part 1*, 1968. ISBN 90 6196 025 8.

MCT 18 R.P. VAN DE RIET, *Formula manipulation in ALGOL 60, part 2*, 1968. ISBN 90 6196 038 X.

MCT 19 J. VAN DER SLOT, *Some properties related to compactness*, 1968. ISBN 90 6196 026 6.

MCT 20 P.J. VAN DER HOUWEN, *Finite difference methods for solving partial differential equations*, 1968. ISBN 90 6196 027 4.

MCT 21 E. WATTEL, *The compactness operator in set theory and topology*, 1968. ISBN 90 6196 028 2.

MCT 22 T.J. DEKKER, *ALGOL 60 procedures in numerical algebra, part 1*, 1968. ISBN 90 6196 029 0.

MCT 23 T.J. DEKKER & W. HOFFMANN, *ALGOL 60 procedures in numerical algebra, part 2*, 1968. ISBN 90 6196 030 4.

MCT 24 J.W. DE BAKKER, *Recursive procedures*, 1971. ISBN 90 6196 060 6.

MCT 25 E.R. PAERL, *Representations of the Lorentz group and projective geometry*, 1969. ISBN 90 6196 039 8.

MCT 26 EUROPEAN MEETING 1968, *Selected statistical papers, part I*, 1968. ISBN 90 6196 031 2.

MCT 27 EUROPEAN MEETING 1968, *Selected statistical papers, part II*, 1969. ISBN 90 6196 040 1.

MCT 28 J. OOSTERHOFF, *Combination of one-sided statistical tests*, 1969. ISBN 90 6196 041 X.

MCT 29 J. VERHOEFF, *Error detecting decimal codes*, 1969. ISBN 90 6196 042 8.

MCT 30 H. BRANDT CORSTIUS, *Excercises in computational linguistics*, 1970. ISBN 90 6196 052 5.

MCT 31 W. MOLENAAR, *Approximations to the Poisson, binomial and hypergeometric distribution functions*, 1970. ISBN 90 6196 053 3.

MCT 32 L. DE HAAN, *On regular variation and its application to the weak convergence of sample extremes*, 1970. ISBN 90 6196 054 1.

MCT 33 F.W. STEUTEL, *Preservation of infinite divisibility under mixing and related topics*, 1970. ISBN 90 6196 061 4.

MCT 34 I. JUHÁSZ, A. VERBEEK & N.S. KROONENBERG, *Cardinal functions in topology*, 1971. ISBN 90 6196 062 2.

MCT 35 M.H. VAN EMDEN, *An analysis of complexity*, 1971. ISBN 90 6196 063 0.

MCT 36 J. GRASMAN, *On the birth of boundary layers*, 1971. ISBN 90 6196 064 9.

MCT 37 J.W. DE BAKKER, G.A. BLAAUW, A.J.W. DUIJVESTIJN, E.W. DIJKSTRA, P.J. VAN DER HOUWEN, G.A.M. KAMSTEEG-KEMPER, F.E.J. KRUSEMAN ARETZ, W.L. VAN DER POEL, J.P. SCHAAP-KRUSEMAN, M.V. WILKES & G. ZOUTENDIJK, *MC-25 Informatica Symposium*, 1971. ISBN 90 6196 065 7.

MCT 38 W.A. VERLOREN VAN THEMAAT, *Automatic analysis of Dutch compound words*, 1971. ISBN 90 6196 073 8.

MCT 39 H. BAVINCK, *Jacobi series and approximation*, 1972. ISBN 90 6196 074 6.

MCT 40 H.C. TIJMS, *Analysis of (s,S) inventory models*, 1972. ISBN 90 6196 075 4.

MCT 41 A. VERBEEK, *Superextensions of topological spaces*, 1972. ISBN 90 6196 076 2.

MCT 42 W. VERVAAT, *Success epochs in Bernoulli trials (with applications in number theory)*, 1972. ISBN 90 6196 077 0.

MCT 43 F.H. RUYMGAART, *Asymptotic theory of rank tests for independence*, 1973. ISBN 90 6196 081 9.

MCT 44 H. BART, *Meromorphic operator valued functions*, 1973. ISBN 90 6196 082 7.

MCT 45 A.A. BALKEMA, *Monotone transformations and limit laws*, 1973. ISBN 90 6196 083 5.

MCT 46 R.P. VAN DE RIET, *ABC ALGOL, A portable language for formula manipulation systems, part 1: The language*, 1973. ISBN 90 6196 084 3.

MCT 47 R.P. VAN DE RIET, *ABC ALGOL, A portable language for formula manipulation systems, part 2: The compiler*, 1973. ISBN 90 6196 085 1.

MCT 48 F.E.J. KRUSEMAN ARETZ, P.J.W. TEN HAGEN & H.L. OUDSHOORN, *An ALGOL 60 compiler in ALGOL 60, Text of the MC-compiler for the EL-X8*, 1973. ISBN 90 6196 086 X.

MCT 49 H. KOK, *Connected orderable spaces*, 1974. ISBN 90 6196 088 6.

MCT 50 A. VAN WIJNGAARDEN, B.J. MAILLOUX, J.E.L. PECK, C.H.A. KOSTER, M. SINTZOFF, C.H. LINDSEY, L.G.L.T. MEERTENS & R.G. FISKER (Eds), *Revised report on the algorithmic language ALGOL 68*, 1976. ISBN 90 6196 089 4.

MCT 51 A. HORDIJK, *Dynamic programming and Markov potential theory*, 1974. ISBN 90 6196 095 9.

MCT 52 P.C. BAAYEN (ed.), *Topological structures*, 1974. ISBN 90 6196 096 7.

MCT 53 M.J. FABER, *Metrizability in generalized ordered spaces*, 1974. ISBN 90 6196 097 5.

MCT 54 H.A. LAUWERIER, *Asymptotic analysis, part 1*, 1974. ISBN 90 6196 098 3.

MCT 55 M. HALL JR. & J.H. VAN LINT (Eds), *Combinatorics, part 1: Theory of designs, finite geometry and coding theory*, 1974. ISBN 90 6196 099 1.

MCT 56 M. HALL JR. & J.H. VAN LINT (Eds), *Combinatorics, part 2: graph theory, foundations, partitions and combinatorial geometry*, 1974. ISBN 90 6196 100 9.

MCT 57 M. HALL JR. & J.H. VAN LINT (Eds), *Combinatorics, part 3: Combinatorial group theory*, 1974. ISBN 90 6196 101 7.

MCT 58 W. ALBERS, *Asymptotic expansions and the deficiency concept in statistics*, 1975. ISBN 90 6196 102 5.

MCT 59 J.L. MIJNHEER, *Sample path properties of stable processes*, 1975. ISBN 90 6196 107 6.

MCT 60 F. GÖBEL, *Queueing models involving buffers*, 1975. ISBN 90 6196 108 4.

* MCT 61 P. VAN EMDE BOAS, *Abstract resource-bound classes, part 1*. ISBN 90 6196 109 2.

* MCT 62 P. VAN EMDE BOAS, *Abstract resource-bound classes, part 2*. ISBN 90 6196 110 6.

MCT 63 J.W. DE BAKKER (ed.), *Foundations of computer science*, 1975. ISBN 90 6196 111 4.

MCT 64 W.J. DE SCHIPPER, *Symmetric closed categories*, 1975. ISBN 90 6196 112 2.

MCT 65 J. DE VRIES, *Topological transformation groups 1 A categorical approach*, 1975. ISBN 90 6196 113 0.

MCT 66 H.G.J. PIJLS, *Locally convex algebras in spectral theory and eigenfunction expansions*, 1976. ISBN 90 6196 114 9.

* MCT 67 H.A. LAUWERIER, *Asymptotic analysis, part 2.* ISBN 90 6196 119 X.

MCT 68 P.P.N. DE GROEN, *Singularly perturbed differential operators of second order*, 1976. ISBN 90 6196 120 3.

MCT 69 J.K. LENSTRA, *Sequencing by enumerative methods*, 1977. ISBN 90 6196 125 4.

MCT 70 W.P. DE ROEVER JR., *Recursive program schemes: semantics and proof theory*, 1976. ISBN 90 6196 127 0.

MCT 71 J.A.E.E. VAN NUNEN, *Contracting Markov decision processes*, 1976. ISBN 90 6196 129 7.

MCT 72 J.K.M. JANSEN, *Simple periodic and nonperiodic Lamé functions and their applications in the theory of conical waveguides*, 1977. ISBN 90 6196 130 0.

MCT 73 D.M.R. LEIVANT, *Absoluteness of intuitionistic logic*, 1979. ISBN 90 6196 122 x.

MCT 74 H.J.J. TE RIELE, *A theoretical and computational study of generalized aliquot sequences*, 1976. ISBN 90 6196 131 9.

MCT 75 A.E. BROUWER, *Treelike spaces and related connected topological spaces*, 1977. ISBN 90 6196 132 7.

MCT 76 M. REM, *Associons and the closure statement*, 1976. ISBN 90 6196 135 1.

MCT 77 W.C.M. KALLENBERG, *Asymptotic optimality of likelihood ratio tests in exponential families*, 1977 ISBN 90 6196 134 3.

MCT 78 E. DE JONGE, A.C.M. VAN ROOIJ, *Introduction to Riesz spaces*, 1977. ISBN 90 6196 133 5.

MCT 79 M.C.A. VAN ZUIJLEN, *Empirical distributions and rankstatistics*, 1977. ISBN 90 6196 145 9.

MCT 80 P.W. HEMKER, *A numerical study of stiff two-point boundary problems*, 1977. ISBN 90 6196 146 7.

MCT 81 K.R. APT & J.W. DE BAKKER (Eds), *Foundations of computer science II*, part 1, 1976. ISBN 90 6196 140 8.

MCT 82 K.R. APT & J.W. DE BAKKER (Eds), *Foundations of computer science II*, part 2, 1976. ISBN 90 6196 141 6.

MCT 83 L.S. VAN BENTEM JUTTING, *Checking Landau's "Grundlagen" in the AUTOMATH system*, 1979 ISBN 90 6196 147 5.

MCT 84 H.L.L. BUSARD, *The translation of the elements of Euclid from the Arabic into Latin by Hermann of Carinthia (?) books vii-xii*, 1977. ISBN 90 6196 148 3.

MCT 85 J. VAN MILL, *Supercompactness and Wallman spaces*, 1977. ISBN 90 6196 151 3.

MCT 86 S.G. VAN DER MEULEN & M. VELDHORST, *Torrix I*, 1978. ISBN 90 6196 152 1.

* MCT 87 S.G. VAN DER MEULEN & M. VELDHORST, *Torrix II*, ISBN 90 6196 153 x.

MCT 88 A. SCHRIJVER, *Matroids and linking systems*, 1977. ISBN 90 6196 154 8.

MCT 89 J.W. DE ROEVER, *Complex Fourier transformation and analytic functionals with unbounded carriers*, 1978. ISBN 90 6196 155 6.

* MCT 90 L.P.J. GROENEWEGEN, *Characterization of optimal strategies in dynamic games*, . ISBN 90 6196 156 4.

MCT 91 J.M. GEYSEL, *Transcendence in fields of positive characteristic*, 1979. ISBN 90 6196 157 2.

MCT 92 P.J. WEEDA, *Finite generalized Markov programming*, 1979. ISBN 90 6196 158 0.

MCT 93 H.C. TIJMS (ed.) & J. WESSELS (ed.), *Markov decision theory*, 1977. ISBN 90 6196 160 2.

MCT 94 A. BIJLSMA, *Simultaneous approximations in transcendental number theory*, 1978. ISBN 90 6196 162 9.

MCT 95 K.M. VAN HEE, *Bayesian control of Markov chains*, 1978. ISBN 90 6196 163 7.

* MCT 96 P.M.B. VITÁNYI, *Lindenmayer systems: structure, languages, and growth functions*, . ISBN 90 6196 164 5.

* MCT 97 A. FEDERGRUEN, *Markovian control problems; functional equations and algorithms*, . ISBN 90 6196 165 3.

MCT 98 R. GEEL, *Singular perturbations of hyperbolic type*, 1978. ISBN 90 6196 166 1

MCT 99 J.K. LENSTRA, A.H.G. RINNOOY KAN & P. VAN EMDE BOAS, *Interfaces between computer science and operations research*, 1978. ISBN 90 6196 170 X.

MCT 100 P.C. BAAYEN, D. VAN DULST & J. OOSTERHOFF (Eds), *Proceedings bicentennial congress of the Wiskundig Genootschap, part 1*, 1979. ISBN 90 6196 168 8.

MCT 101 P.C. BAAYEN, D. VAN DULST & J. OOSTERHOFF (Eds), *Proceedings bicentennial congress of the Wiskundig Genootschap, part 2*, 1979. ISBN 90 9196 169 6.

MCT 102 D. VAN DULST, *Reflexive and superreflexive Banach spaces*, 1978. ISBN 90 6196 171 8.

MCT 103 K. VAN HARN, *Classifying infinitely divisible distributions by functional equations*, 1978. ISBN 90 6196 172 6.

MCT 104 J.M. VAN WOUWE, *Go-spaces and generalizations of metrizability*, 1979. ISBN 90 6196 173 4.

* MCT 105 R. HELMERS, *Edgeworth expansions for linear combinations of order statistics*, . ISBN 90 6196 174 2.

MCT 106 A. SCHRIJVER (Ed.), *Packing and covering in combinatorics*, 1979. ISBN 90 6196 180 7.

MCT 107 C. DEN HEIJER, *The numerical solution of nonlinear operator equations by imbedding methods*, 1979. ISBN 90 6196 175 0.

MCT 108 J.W. DE BAKKER & J. VAN LEEUWEN (Eds), *Foundations of computer science III*, part 1, 1979. ISBN 90 6196 176 9.

MCT 109 J.W. DE BAKKER & J. VAN LEEUWEN (Eds), *Foundations of computer science III*, part 2, 1979. ISBN 90 6196 177 7.

MCT 110 J.C. VAN VLIET, *ALGOL 68 transput*, part I: *Historical Review and Discussion of the Implementation Model*, 1979. ISBN 90 6196 178 5.

MCT 111 J.C. VAN VLIET, *ALGOL 68 transput*, part II: *An implementation model*, 1979. ISBN 90 6196 179 3.

MCT 112 H.C.P. BERBEE, *Random walks with stationary increments and Renewal theory*, 1979. ISBN 90 6196 182 3.

MCT 113 T.A.B. SNIJDERS, *Asymptotic optimality theory for testing problems with restricted alternatives*, 1979. ISBN 90 6196 183 1.

MCT 114 A.J.E.M. JANSSEN, *Application of the Wigner distribution to harmonic analysis of generalized stochastic processes*, 1979. ISBN 90 6196 184 x.

MCT 115 P.C. BAAYEN & J. VAN MILL (Eds), *Topological Structures II*, part 1, 1979. ISBN 90 6196 185 5.

MCT 116 P.C. BAAYEN & J. VAN MILL (Eds), *Topological Structures II*, part 2, 1979. ISBN 90 6196 186 6.

MCT 117 P.J.M. KALLENBERG, *Branching processes with continuous state space*, 1979. ISBN 90 6196 188 2.

MCT 118 P. GROENEBOOM, *Large deviations and Asymptotic efficiencies*, 1980. ISBN 90 6196 190 4.

MCT 119 F. PETERS, *Sparse matrices and substructures*, 1980. ISBN 90 6196 192 0.

MCT 120 W.P.M. DE RUYTER, *On the Asymptotic Analysis of Large Scale Ocean Circulation*, 1980. ISBN 90 6196 192 9.

MCT 121 W.H. HAEMERS, *Eigenvalue techniques in design and graph theory*, 1980. ISBN 90 6196 194 7.

MCT 122 J.C.P. BUS, *Numerical solution of systems of nonlinear equations*, 1980. ISBN 90 6196 195 5.

MCT 123 I. YUHÁSZ, *Cardinal functions intopology - ten years later*, 1980. ISBN 90 6196 196 3.

MCT 124 R.D. GILL, *Censoring and Stochastic Integrals*, 1980. ISBN 90 6196 197 1.

AN ASTERISK BEFORE THE NUMBER MEANS "TO APPEAR"

Addendum to

MAY 1974

SOCIAL SERVICES IN BRITAIN

Since this pamphlet was written a major reorganisation of local government, health services and water supply and sewerage has taken effect,[1] and there has also been a change of Government, leading to the adoption of new social policies. There have, accordingly, been a number of major developments, primarily and immediately in social security and health services, though changes in education and housing policy may lead fairly rapidly to changes in the pattern of services.

SOCIAL SECURITY

From 21 January 1974, the flat-rate weekly contributions of employers were raised to £1·337 and £1·131 in respect of men and women contracted out, and to £1·217 and £1·051 in respect of men and women not contracted out, while the rates for self-employed rose to £1·99 for men and £1·67 for women and for non-employed to £1·56 for men and £1·23 for women.

Following the change of Government on 4 March 1974, the new Secretary of State for Social Services, Mrs Barbara Castle, announced further large increases in benefits and changes in contributions to be introduced on 22 July 1974, together with a relaxation of the earnings rule, so that pensioners can earn up to £13 a week without reduction of pension. In addition the annual review of social security benefits will in future seek to keep them in line with increases in the general level of wages rather than the cost of living.

The standard long-term flat rate of national insurance benefits (for invalidity, widows and retirement pensions and widowed mothers' allowance), will, from 22 July 1974, become £10 a week for a single, widowed or divorced person or a married man and £6 for a wife or adult dependant, while the standard short-term benefits (for unemployment and sickness) will become £8·60 for a single, widowed or divorced person or a married man and £5·30 for a wife or adult dependant. Other social security benefits have been increased—usually more or less in proportion. Details are given in the tables included in the recently issued Addendum No. 2 to R5455/73, *Social Security in Britain.*

Flat-rate contributions, including industrial injuries and National Health Service contributions, will, on 5 August 1974, go down to £0·87 and £0·70 respectively for employed men and women contracted out, and £0·75 and £0·62 respectively if not contracted out, while the corresponding employers' contributions will rise to £1·777, £1·511, £1·657 and £1·431. Contributions of the self-employed will rise to £2·41 for men and £2·01 for women, and those of the non-employed to £1·90 for men and £1·49 for women.

The methods of calculation of graduated (earnings-related) benefits will remain unchanged. The rate of earnings-related contributions, however, is to be raised to 5·5 per cent on the whole range from £9 to £62 for employees not contracted out, while for those contracted out, the rate is to be 1·25 per cent on the range from £9 to £18 and 5·5 per cent on the range from £18 to £62.

Mrs Castle also reaffirmed the Government's intention to replace the Social Security Act 1973 and, pending its replacement, to allow only some of its provisions, those relating to the basic flat-rate pension financed by earnings-related contributions and to the preservation of pension rights, to come into force in April 1975. The provisions for a state reserve scheme with earnings-related contributions and benefits and for exemption from this scheme for approved occupational schemes would not come into force.

---

[1]See COI short note, *Reorganisation of Local Government, Water and Health Services* SN5967.

HEALTH SERVICES

The reorganisation of health services, of local government and water supply and sewerage administration came into effect in England and Wales on 1 April 1974, at the same time as the reorganisation of health services in Scotland, where reorganisation of local government is not due until May 1975.

The reorganisation of health services has taken place on the lines of the proposals summarised on page 24 of the pamphlet, but the Secretary of State for Social Services in the new Government has announced her intention of introducing, within the framework of the existing legislation, certain modifications in the new health service administration in England, in order to make it more representative of local interests. Family planning advice and help have been made freely available to all (without a prescription charge for drugs and appliances) at hospitals, clinics and centres. Negotiations with GPs continue.

Public Health

Local authorities continue to have responsibilities and powers in relation to public health. They share, however, many responsibilities with the area health authorities (health boards in Scotland), and for this reason they are recommended to nominate a medical officer of the area health authority as their 'proper officer' to undertake such duties. At the same time the main responsibility for water supply and sewerage and sewage disposal has passed in England and Wales to the new regional water authorities, though they may use local authorities as their agents in respect of sewerage and existing water undertakings in respect of water supply.

EDUCATION

The new Government's policy is to encourage the development of comprehensive secondary schools, in order to put an end as soon as possible to selection at the age of 11.

HOUSING

The Government has imposed a freeze on all residential rents during 1974. It has also announced its intention of encouraging local authority building and acquisition of residential property to rent, and of introducing legislation to give to tenants of furnished accommodation greater security of tenure, similar to that enjoyed by tenants of unfurnished accommodation. The Housing Bill at present before Parliament contains measures to deal with areas of housing stress, to make the improvement grant system more effective and to expand the role of the voluntary housing movement.

Prepared by Reference Division, Central Office of Information, London.

May 1974.

London: Her Majesty's Stationery Office

Printed in England for Her Majesty's Stationery Office by Walter Jenn Ltd., London,

Dd. 098710 K28 5/74